SOTHIS PRESS

Sothis Press aims to completely reimagine the public domain book with an emphasis on universal access to captivatingly designed print editions. We believe that technology will enable every single book ever published to be perpetually accessible in beautiful print form. Sothis is beauty in ubiquity.

www.sothis.press
1 Waterhouse Square, London, ECIN 2ST

The Carpenter's New Guide

A Complete Book of Lines for Carpentry and Joinery; Treating Fully on Practical Geometry, Soffits, Groins, Niches, Roofs, and Domes, and Containing a Great Variety of Original Designs, Also a Full Exemplification of the Theory and Practice of Every Descri

Peter Nicholson

SOTHIS PRESS

ISBN 978-1-0365-2092-2

The Carpenter's New Guide
is in the public domain.

This edition is published by
 The company
is registered in England and Wales.
The Company Number is 14900389.

THE

CARPENTER'S NEW GUIDE:

BEING A

COMPLETE BOOK OF LINES

FOR

CARPENTRY AND JOINERY.

TREATING FULLY ON

Practical Geometry,

SOFFITS, BRICK AND PLASTER GROINS, NICHES OF EVERY DESCRIPTION, SKY-LIGHTS, LINES FOR ROOFS AND DOMES;

WITH A GREAT VARIETY OF

DESIGNS FOR ROOFS, TRUSSED GIRDERS, FLOORS, DOMES, BRIDGES, &c.; ANGLE BARS FOR SHOP FRONTS, &c.; AND RAKING MOULDINGS.

ALSO

ADDITIONAL PLANS FOR VARIOUS STAIR-CASES, WITH THE LINES FOR PRODUCING THE FACE AND FALLING MOULDS NEVER BEFORE PUBLISHED, AND GREATLY SUPERIOR TO THOSE GIVEN IN THE FORMER EDITION OF THIS WORK,

BY WILLIAM JOHNSTON, ARCHITECT,
OF PHILADELPHIA.

THE WHOLE FOUNDED ON TRUE GEOMETRICAL PRINCIPLES;

THE THEORY AND PRACTICE WELL EXPLAINED AND FULLY EXEMPLIFIED ON

EIGHTY-THREE COPPER-PLATES:

INCLUDING

SOME OBSERVATIONS AND CALCULATIONS ON THE STRENGTH OF TIMBER.

BY PETER NICHOLSON,
AUTHOR OF "THE CARPENTER AND JOINER'S ASSISTANT," "THE STUDENT'S INSTRUCTOR TO THE FIVE ORDERS," ETC.

THIRTEENTH EDITION.

PHILADELPHIA:
GRIGG, ELLIOT AND CO., No. 14 NORTH FOURTH STREET.
1848.

PHILADELPHIA:
T. K. AND P. G. COLLINS, PRINTERS.

PREFACE.

To a book intended merely for the use of Practical Mechanics, much Preface is not necessary:—It is proper, however, to say, that whatever rules by previous authors have on examination proved to be true and well explained, these have been selected and adopted, with such alterations as a very close attention has warranted for the more easily comprehending them, for their greater accuracy or facility of application; added to these, are many examples which are entirely of my own invention, and such as will, I am persuaded, conduce very much to the accuracy of the work and to the ease of the workman.

The arrangement of the subjects in this work is gradual and regular, and such as a student should pursue who wishes to attain a thorough knowledge of his profession: and as it is Geometry that lays down all the first principles of building, measures of lines, angles, and solids, and gives rules for describing the various kinds of figures used in buildings; therefore, as a necessary introduction to the art treated of, I have first laid down, and explained in the terms of workmen, such problems of GEOMETRY as are absolutely requisite to the well understanding and putting in practice the necessary lines for Carpentry. These problems duly considered, and their results well understood, the learner may proceed to the theoretical part of the subject, in which Soffits claim particular attention; for, by a thorough knowledge of these, the student will be enabled to lay down arches which shall stand exactly perpendicular over their plan, whatever form the plan may be: on this depends the well executing all groins, arches, niches, &c., constructed in circular walls, or which stand upon irregular bases; wherefore the importance of rightly understanding these I cannot too much insist upon, their construction being so various and intricate, and their uses so frequently required. The two plates of cuneoidal or winding soffits are new, and are constructed in a more simple and more accurate manner: yet this method is only a nearer approximation to truth than the former one; the surface of a conchoid cannot be developed; that is, it cannot be extended on a plane: it is therefore absurd to look for perfection on this subject.

The next subject which regularly presents itself is *Groins*; for the construction of which there will be found many methods entirely new; and besides the common figures, I have shown many which are difficult of execution, and not to be found in any other author. I have displayed a variety of methods for constructing spherical niches, a form more frequently wanted than the elliptic, which only has yet been explained.

Among the various methods for finding the Lines for Roofs, I have given an entire new one for finding the down and side bevels of purlines, so that they shall exactly fit against the hip rafter; and by the same method the jack rafter will be made to fit.

Of Domes and Polygons, I have shown an entirely new method for finding their covering, within the space of the board, thereby avoiding the tedious and incommodious method of finding the lines on the dome itself, as has been always practised heretofore: also a method for finding the form of the boards near the bottom, when a dome is to be covered horizontally. Of dome-lights over stair-cases, or in the centre of groins, a rule upon true principles is given, for finding their proper curve against the wall, and the curve of the ribs; this has never before been made public.

Having gone thus far in the Art of Carpentry, it is necessary for me to say, by way of caution and guard to the ardent theorist, that there are some surfaces which cannot be developed; such as spherical or superoidical domes, where their coverings cannot be found by any other means than by supposing the curved surface to become polygonal; in which case such domes may be covered upon true principles, as may be demonstrated. Let us suppose a polygonal dome inscribed in a spherical one; then, the greater the number of sides of the polygonal dome, the nearer it will coincide with its circumscribing spherical one. Again, let us suppose that this polygonal dome has an infinite number of sides; then, its surface will exactly coincide with the spherical dome, and therefore in anything which we shall have occasion to practice, this method will be sufficiently near; as, for example, in a dome of one hundred sides, of a foot each, the rule for finding such a covering will give the practice so very near, that the variation from absolute truth could not be perceived.

PREFACE.

Having gone through the constructive part of Carpentry, I next proceed to examples showing the best forms of floors, partitions, trusses for roofs, truss girders, domes, &c., which shall resist their own weight, or the addition of any adventitious load.

To conclude: as I pretend not to infallibility, I hope to be judged with candor, being always open to conviction, from a knowledge of the difficulty and intricacy of science; yet I hope that my labors may be of some use to others in shortening the road, and smoothing the path through which, for many years, I have been a persevering traveler for knowledge: I shall then be satisfied, and not deem my time misspent if my labors tend to the public good.

P. NICHOLSON.

CONTENTS.

PRACTICAL GEOMETRY.

CARPENTRY.

JOINERY.

CONCLUSION.

PRACTICAL GEOMETRY.

GEOMETRY is the science of extension and magnitude; it teaches the construction of all right-lined and curvilineal figures, and is divided into Theoretical and Practical.

The Theoretical part is founded upon reason and self-evidence: it demonstrates the construction of variously formed figures, and evinces the truth, or detects the falsehood on which they are made. This is the foundation of the Practical part; and without a knowledge of the Theory, no invention to any degree certain can be made in the Practice.

The uses of Geometry are not confined to Carpentry and Architecture, but in the various branches of the Mathematics, it opens and discovers to us their secrets. It teaches us to contemplate truths, to trace the chain of them, subtile and almost imperceptible as it frequently is, and to follow them to the utmost extent.

Its uses are great and necessary in Astronomy and Geography. The science of Perspective is entirely dependent upon its principles. To enumerate its many uses is beyond my power. Those who desire to become thoroughly acquainted with Geometry, will do well to study attentively the elements of Euclid.

As my labors are not intended for the abstruse Mathematician, but for the instruction of the *Practical Carpenter*, I shall omit all speculative demonstrations, the sections of Cylinders and Globes excepted (which are not to be found in Euclid), and confine myself to the useful part of the science, viz. PRACTICAL GEOMETRY.

2

PLATE 1.

DEFINITIONS.

1. *A POINT has position but not magnitude.*

2. *A line is length without breadth or thickness.*

3. *A superficies hath length and breadth only.*

4. *A solid is a figure of three dimensions, having length, breadth, and thickness.*

Hence surfaces are the extremities of solids, and lines the extremities of surfaces, and points the extremities of lines.

5. *Lines are either right, curved, or mixed of these two.*

6. *A right or straight line lies all in the same direction between its extremities, and is the shortest distance between two points, as* A.

7. *A curve continually changes its directions between the extreme points, as* C.

8. *Lines are either parallel, oblique, perpendicular, or tangential.*

9. *Parallel lines are always at the same distance, and will never meet, though ever so far produced, as* D *and* E.

10. *Oblique right lines change their distance, and would meet, if produced, as* F.

11. *One line is perpendicular to another when it inclines no more to one side than another, as* G.

12. *A straight line is a tangent to a curve when it is produced and touches it without cutting, as* H.

13. *An angle is the inclination of two lines towards one another in the same plane, meeting in a point, as* I.

14. *Angles are either right, acute, or oblique, as* K.

15. *A right angle is that which is made by one line perpendicular to another, or when the angles on each side are equal, as* G.

16. *An acute angle is less than a right angle, as* K.

17. *An obtuse angle is greater than a right angle, as* L.

18. *A superficies is either plane or curved.*

19. *A plane, or plane surface is that to which a right line will every way coincide;—but if not, it is curved.*

20. *Plane figures are bounded either by right lines or curves.*

21. *A solid is said to be cut by a plane when it is cut through in any particular place, and the place that is cut is called the section of the solid.*

22. *Plane figures, bounded by right lines, have names according to the number of their sides, or of their angles, for they have as many sides as angles—the least number is three.*

23. *An equilateral triangle is that whose three sides are equal, as* M.

24. *An isosceles triangle has only two sides equal, as* N.

25. *A scalene triangle has all sides unequal, as* O.

26. *A right-angled triangle has one right angle, as* P.

27. *Other triangles are oblique angled, and are either obtuse or acute.*

28. *An acute-angled triangle has all its angles acute, as* M or N.

29. *An obtuse-angled triangle has one obtuse angle, as* O.

30. *A figure of four sides and angles is called a quadrangle, or quadrilateral, as* Q, R, S, T, U, *and* V.

Plate 1.

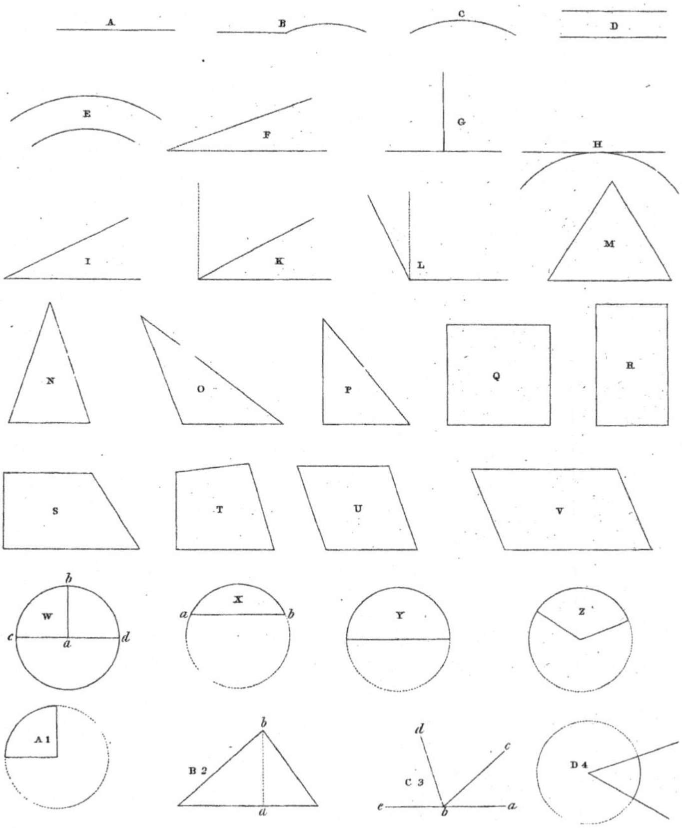

31. *A parallelogram is a quadrilateral, which has both pairs of its opposite sides parallel, as* Q, R, U, *and* V; *and takes the following particular names.*

32. *A rectangle is a parallelogram having all its angles right ones, as* Q *and* R.

33. *A square is an equilateral rectangle, having all its sides equal, and all its angles right ones, as* Q.

34. *A rhombus is an equilateral parallelogram, whose angles are oblique, as* U.

35. *A rhomboid is an oblique-angled parallelogram, as* V.

36. *A trapezium is a quadrilateral which has neither pair of its sides parallel, as* T.

37. *A trapezoid hath only one pair of its opposite sides parallel, as* S.

38. *Plane figures having more than four sides are in general called polygons, and receive other particular names according to the number of their sides or angles.*

39. *A pentagon is a polygon of five sides, a hexagon has six sides, a heptagon seven, an octagon eight, a nonagon nine, a decagon ten, an undecagon eleven, and a dadecagon twelve sides.*

40. *A regular polygon has all its sides and its angles equal; and if they are not equal, the polygon is irregular.*

41. *An equilateral triangle is also a regular figure of three sides, and a square is one of four; the former being called a trigon, and the latter a tetragon.*

42. *A circle is a plane figure bounded by a curve line called the circumference, which is everywhere equidistant from a certain point within, called its centre.*

43. *The radius of a circle is a right line drawn from the centre to the circumference, as* a b *at* W.

44. *A diameter of a circle is a right line drawn through the centre, terminating on both sides of the circumference, as* c d *at* W.

45. *An arch of a circle is any part of the circumference.*

46. *A chord is a right line joining the extremities of an arch, as* a b *at* X.

47. *A segment is any part of a circle bounded by an arch and its cord, as* X.

48. *A semicircle is half the circle, or a segment cut off by the diameter, as* Y.

49. *A sector is any part of a circle bounded by an arch and two radii, drawn to its extremities, as* Z.

50. *A quadrant, or quarter of a circle, is a sector having a quarter of the circumference for its arch, and the two radii are perpendicular to each other, as* A 1.

51. *The height or altitude of any figure is a perpendicular let fall from an angle, or its vertex, to the opposite side, called the base, as* a b *at* B 2.

52. *When an angle is denoted by three letters, the middle one is the place of the angle, and the other two denote the sides containing that angle; thus, let* a b c *be the angle at* C 3, b *is the angular point, and* a b, *and* b c *are the two sides containing that angle.*

53. *The measure of any right-lined angle is an arch of any circle contained between the two lines which form the angle, and the angular point being in the centre, as* D 4.

PLATE 2.

PROBLEMS.

Figure 1. *To draw a Perpendicular to a given Point in a Line.*

A B is a line, and *c* a given point; take *a* and *b*, two equal distances on each side of *c*, and with the foot of the compasses in *a* and *b* make an intersection *d*, and draw *d c*, which is the perpendicular.

Fig. 2. *To make a Perpendicular with a Ten Foot Rod.*

Let *a b* be six feet, then take eight feet, and from *a* make an arch at *c b*, and from the point *a* with the distance of ten feet across at *c*, then draw *c b*, which is the perpendicular.

Fig. 3. *To let fall a Perpendicular from a given Point to a Line.*

In the given point *c* make an arch to cross the line in *a* and *b*, and from *a* and *b* make an intersection at *d*, and draw *c d* the perpendicular.

Fig. 4. *To draw a Perpendicular upon the End of a Line.*

Take any point *d* at pleasure above the line, and with the distance *d b* make an arch *a b c*, and draw a line *a d* to cut it at *c*, and draw *c b* the perpendicular.

Fig. 5. *To divide a Line into two equal Parts by a Perpendicular.*

From the extreme points *a* and *b* describe two arches to intersect at *c* and *d*, draw *c d*, which divides the line into two equal parts.

Fig. 6. *To divide any given Angle into two equal Angles.*

Take two equal distances *a b* and *a c* on each side of the angular point *a*, and with the same opening of the compasses or any other, place the foot in *b* and *c*, make an intersection at *d*, and draw *d a*, which will divide the angle into two equal parts.

Figs. 7 and 8. *An Angle being given, to make another equal to it, from a given Point in a right Line.*

Let *b a c* be the angle given, and *c d* a right line, *c* the given point, on *a* make an arch *b c* with any radius, and on *c* with the same radius describe an arch *d e*, take the chord of *b c*, set it from *d* to *e*, and draw *e c*, then the angle *e c d* will be equal to *c a b*.

Plate 2.

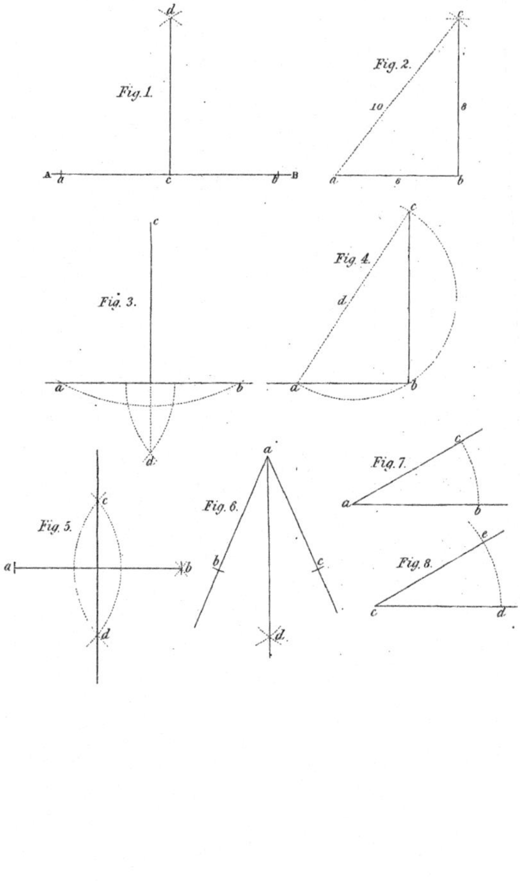
Fig. 1.
d
A
a
c
b
B
Fig. 2.
c
10
8
a
6
b
Fig. 3.
c
a
b
d
Fig. 4.
c
d
a
b
Fig. 5.
c
a
b
d
Fig. 6.
a
b
c
d
Fig. 7.
c
a
b
Fig. 8.
e
c
d

Plate 3.

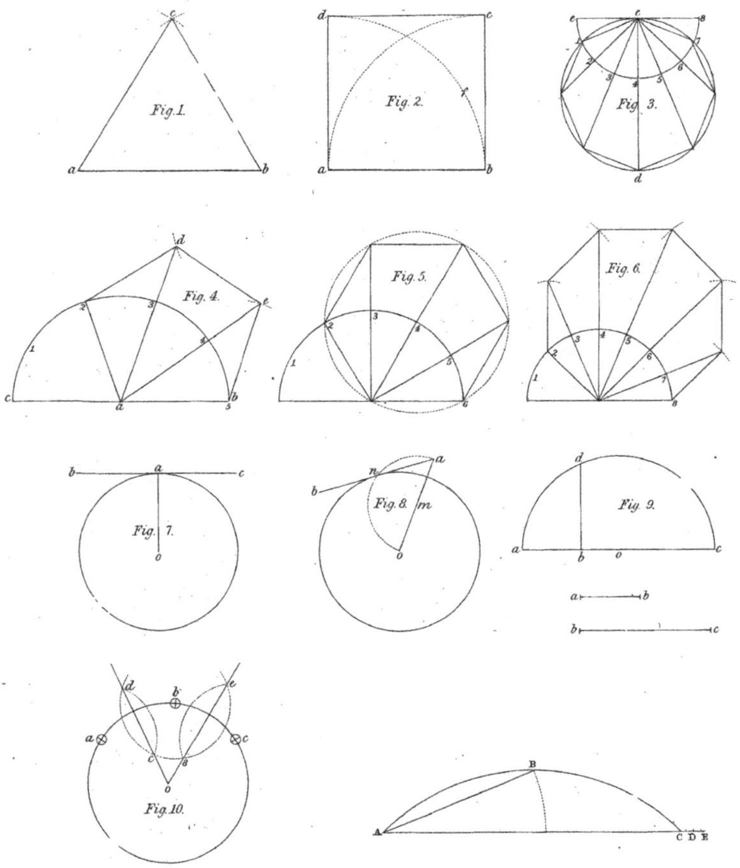

PLATE 3.

Fig. 1. *Upon a right line to make an equilateral Triangle.*

Take *a b* the given side from *a* and *b*, and make an intersection at *c*, and draw *c a* and *c b*.

Fig. 2. *Upon a right Line to make a Square.*

With the given side *a b*, and in the points *a* and *b*, describe two arches to intersect at *e*, divide *b e* into two equal parts at *f*, make *e d* and *e c* each equal to *e f*, draw *a d*, *d c*, and *c b*.

Figs. 4, 5, 6. *The side of any Polygon being given, to describe the Polygon to any Number of sides whatever.*

On one extreme of the given side make a semicircle of any radius, but it will be most convenient to make it equal to the side of the polygon; then divide the semicircle into the same number of equal parts as you would have sides in the polygon, and draw lines from the centre through the divisions in the semicircle, always omitting the two last, and run the given side round each way upon these lines, join each side, and it will be completed.

Example in a Pentagon. Fig. 4.

Let *a b* be the given side, and continue it out to *c*; on *a*, as the centre with the radius *a b*, describe a semicircle, divide it into five equal parts; through 2, 3, 4, draw *a* 2, *a d*, *a e*; make *b e* equal to *a b*, 2 *d* equal to 2 *a* or *a b*; join 2 *d*, *d e*, and *e b*. In the same manner may any other polygon be described.

N. B. This depends upon the equality of the angles upon equal arcs. See Fig. 3.

Fig. 7. *Through a given Point* a, *to draw a Tangent to a given Circle.*

Draw *a o* to the centre, then through *a* draw *b c* perpendicular to *a o*, it will be the tangent.

Fig. 8. *A tangent Line being given, to find the Point where it touches the Circle.*

From any point *a* in the tangent line *b a*, draw a line to the centre *o*, and divide *a o* into two equal parts at *m* and with a radius *m a*, or *m o*, describe an arch, cutting the given circle in *n*, which is the point required.

Fig. 9. *Two right Lines being given, to find a mean Proportion.*

Join *a b* and *b c* in one straight line, divide *a c* into two equal parts at the point *o*, with the radius *o a* or *o c* describe a semicircle, and erect the perpendicular *b d*, then is *a b* to *b d* as *b d* is to *b c*.

Fig. 10. *Through any three Points to describe the Circumference of a Circle.*

From the middle point *b* draw chords *b a* and *b c* to the two other points *a* and *c*, divide the chords *a b* and *b c* into two equal parts by perpendiculars meeting at *O*, which will be the centre.

To find the length of any Arc A B C *of a Circle.*

Draw the chord *A C* and produce it to *E*; bisect the arc *A B C* in *B*, and make *A D* equal to twice *A B*; divide *C D* into three equal parts, and set one out to *E*; then *A E* is the length of the arc.

PLATE 4.

Fig. 1. *Three Lines being given, to form a Triangle.*

Take one of the given sides *a b*, and make it the base of the triangle; take the other side *a c*, and from *a*, describe an arch at *c*; then take the third side *b c*, and from *b*, describe another arch crossing the former at *c*, and join *a c* and *b c*.

Note: that any two lines must be greater than a third.

Figs. 2, 3. *To make a Quadrangle equal to a given Quadrangle.*

Divide the given quadrangle, *fig.* 2, in two triangles; make the triangle *e f g* equal to *a b c*, and *e g h* equal to *a c d*, and it is done.

Figs. 4, 5. *Any irregular Polygon being given, to make another of the same dimensions.*

Divide the given polygon, *fig.* 4, into triangles, and in *fig.* 5, make triangles in the same position, respectively equal to those in *fig.* 4; then will the irregular polygon *f, g, h, i, k,* be equal and similar to *a b c d e*.

Fig. 6. *To make a Rectangle equal to a given Triangle.*

Draw a perpendicular *c d*, divide it into two equal parts at *e*, through *e* draw *f g*, parallel to the base *a b*, draw *a f, b g*, perpendicular; then will the rectangle *a b g f* be equal to the triangle *a b c*.

Fig. 7. *To make a Square equal to a given Rectangle.*

Let *a b c d* be the given rectangle; continue one of its sides as *a b* out to *e*, make *b e* equal to the other side *b c*, divide *a e* into two equal parts at *i*, with the radius *i e* or *i a* make a semicircle *a f e*, and draw *b f* perpendicular to *a b*; make the square *b f g h*, which is equal to the parallelogram *a b c d*.

Fig. 8. *To make a Square equal two to given Squares.*

Make the perpendicular sides *a c* and *a b* of the right-angled triangle *c a b* equal to the sides of the given squares *A* and *B*, draw the hypothenuse *c b*, which is the side of the square *D* equal to the squares *A, B, C*.

Plate 4.

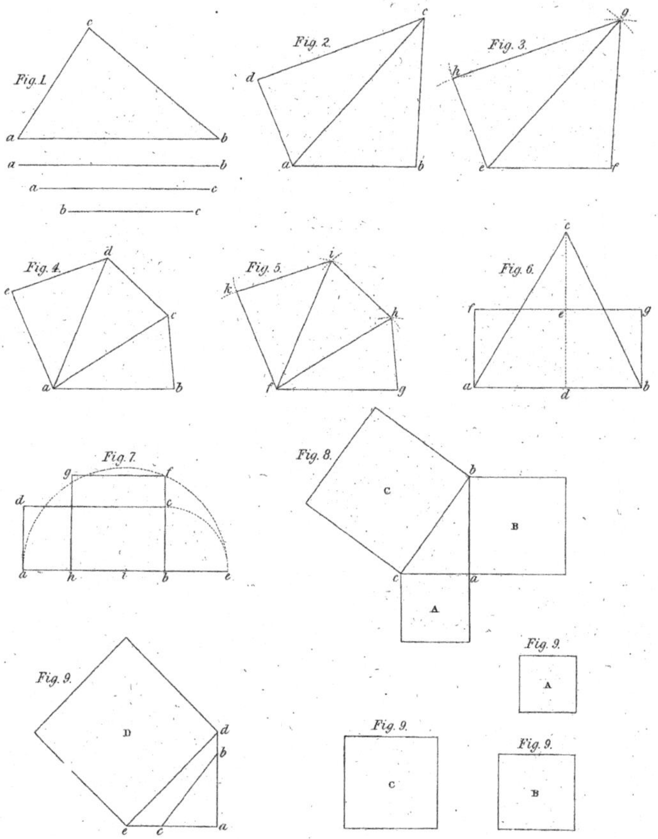

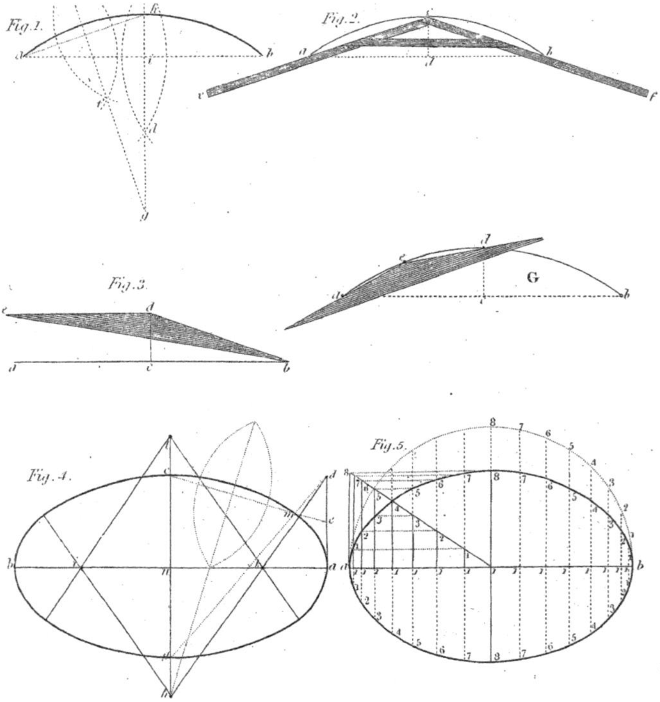
Fig. 1.
Fig. 2.
Fig. 3.
G
Fig. 4.
Fig. 5.

PLATE 5.

Fig. 1. *To draw a Segment of a Circle to any Length and Height.*

a b is the length, *i h* the height; divide the length *a b* into two parts by a perpendicular *g c*; divide *h* by the same method, then their meeting at *g* will be the centre; fix the foot of the compasses in *g*, extend the other leg to *h*, make the arch *a h b*, which is the segment.

Fig. 2. *To draw a Segment by Rods, to any Length and Height.*

Make two rods *c e* and *c f* to form an angle *e c f*, so that each may be equal to *a b*, the opening; place the angle *c* to the height, and the edges to *a* and *b*, put a piece *a b* across them to keep them tight, then move your lath round the points *a*, *b*, and it will describe the segment at the point *c*.

Fig. 3. *To describe a Segment of a Circle at twice, upon true Principles, by a flat Triangle.*

Let the extent of the segment be *a b*, its height *c d*, from the extreme *b* to the top *d* draw *b d*, through the point *d* draw *e d* parallel to the base *a b*, equal in length to *d b*, describe one half, as you see at *G*; then move your nail, or pin, out of *a*, stick it in the point *b*, and describe the other half.

Fig. 4. *The transverse and conjugate Axis of an Ellipsis being given, to draw its Representation.*

Draw *a d* parallel and equal to *n c*, bisect it in *e*; draw *e c* and *d g* cutting each other at *m*, join *m c*, bisect it by a perpendicular meeting *c g*, produced at *h*; draw *h d*, cutting *b a* at *k*, and make *n i* equal to *n k*; *n l* equal to *n h*; through the points *i*, *l*, *k*, *h*, draw the lines *h i*, *k l*, and *i l*, *h k*, then describe the four sectors by help of the centres, *i*, *l*, *k*, *h*, and it will be the representation required.

Fig. 5. *To describe an Ellipsis by Ordinates.*

Make a semicircle on the length *a b*, divide it into any number of equal parts, as 16, on the end at *a* make *a* 8 perpendicular, equal to half the width, and draw the ordinates through all the points in the semicircle, draw the line 8, 1 to the centre, then *a* 1 8 will be a scale to set your oval off; take 1 1 from the scale, and set it from 1 to 1 in your oval both ways at each end; then take 1 2 in your scale, and set it to 1 2 in the oval, and find all the other points in the same manner; a curve being traced through these points will be the true ellipsis.

PLATE 6.

Fig. 1. *To make an Ellipsis with a String.*

Take the half *a g* of the longest diameter *a b*, and with that distance fix the foot of the compass in *c*, cross *a b* at *e f*, in which stick two nails or brads, then lay a string round *e f* and *c*, fix a pencil at *c*, and move your hand round, keeping the string tight, will describe the ellipsis.

Fig. 2. *To describe an Ellipsis by a Trammel.*

1 2 3 is a trammel rod: at 1 is a nut with a hole to hold a pencil; at 2 and 3 are two other sliding nuts; make the distance of 2 from 1, half the shortest diameter of your ellipsis, and from the nut 1 to 3 equal to half the longest, the points 2 and 3 being put into the grooves of the same size, move your pencil round at 1, and it will describe the true curve of an ellipsis.

Fig. 3. *An Ellipsis being given, to find the Centre and two Axes.*

Draw any two parallel lines *a b* and *c d* at pleasure, divide each of them in two equal parts at the points *e* and *f*, and through *e f*, draw the line *k l*, divide *k l* into two equal parts at the point *g*, place the foot of the compass in *g*, with the other foot make two crosses *h* and *i*, on the circumference; draw a line *h i*, through *g*, draw *m n* parallel to *h i*, also through *g* draw *o p*, at right angles to *m n*; then *o p* is the transverse axis, and *m n* the conjugate, and *g* the centre of an ellipsis.

Fig. 4. *How to proportionate one Ellipsis within another; that is, to give it the same Length in Proportion to its Width, as the Length of the other has to its Width.*

Let the given ellipsis be *a d b c*, make the parallelogram *e h f g* to touch the sides and ends of the ellipsis, draw the diagonals *e f*, and *g h*, of the rectangle, let *r q* be the width of the lesser ellipsis given, through the point *q*, or *r*, draw *l o*, or *m n*, parallel to the transverse axis, at the points *m* and *n*, where it cuts the diagonal, draw *m l* and *n o* parallel to the conjugate axis, will also show its length.

Fig. 5. *How to describe an Ellipsis about a Parallelogram, to have the same Length in Proportion to its Width, as the Length of the Parallelogram has to its Width.*

Let the given parallelogram be *a b c d*; let the diagonals *a c*, and *b d*, be drawn from the centre *i*; draw the quarter of a circle, 2 1 *k*, to half the width of a rectangle; divide the quadrant into two equal parts at 1; through the point 1, draw the line *l* 3 parallel to the transverse axis to cut the diagonal *b d* in the point 3; then draw the lines 3 2 and 3 4; again, draw *f d* parallel to 2 3, then *i f* will be half the width, and *d e* parallel to 3 4; and *i e* will be half the length of the ellipsis: make *i h* equal to *i e*, and *i g* equal to *i f*, which will give the four points through which the ellipsis must pass; describe the curve, and the thing will be done.

Fig. 6. *To divide a Line in the same Proportion as another is divided.*

d a is a line given already divided, and *d e* is a line to be divided in the same proportion, making any angle at *d* join *a e*, draw *b f* and *c g* parallel to *a e*; then *d e* is divided at *f* and *g* in the ratio of *a d* at *b* and *c*.

Fig. 7. *To do the same by an equilateral Triangle.*

a b is the given line divided: from *c* take two equal distances *c d*, *e d*, and by drawing lines from the several points in *a b* to *c* cutting *d d*, *d d* will be divided as *a b*.

Fig. 8. *To make an Octagon the nearest Way from a Square.*

Draw the diagonals of the square to cross at *e*, fix the foot of your compass in *c*, and take the distance *c e* and make an arch *f e g*; then set your gauge to *d f* or *b g*, which will gauge off each angle.

Plate 6.

Fig. 1.

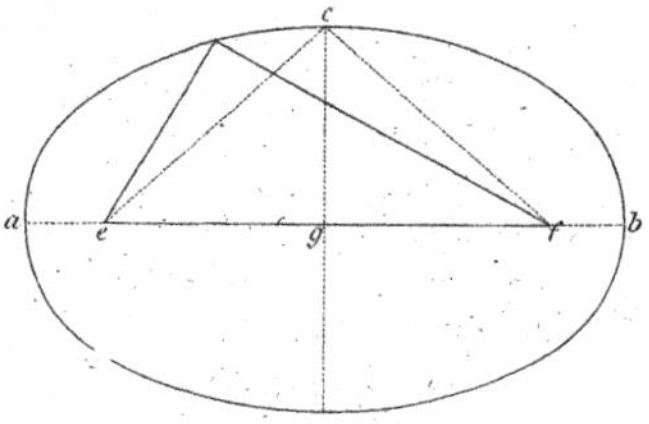

Fig. 2.

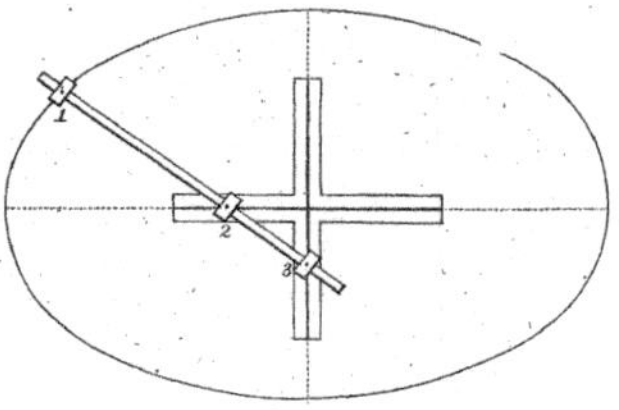

Fig. 3.

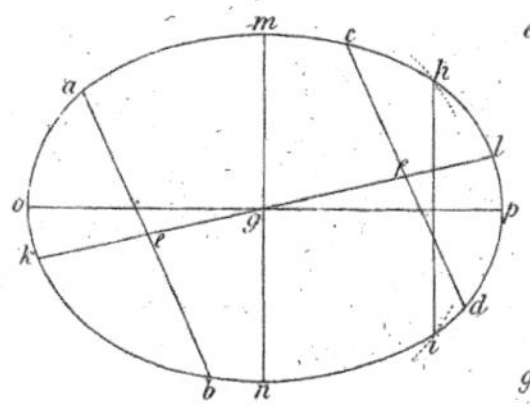

Fig. 4.

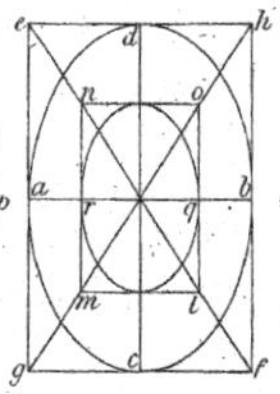

Fig. 5.

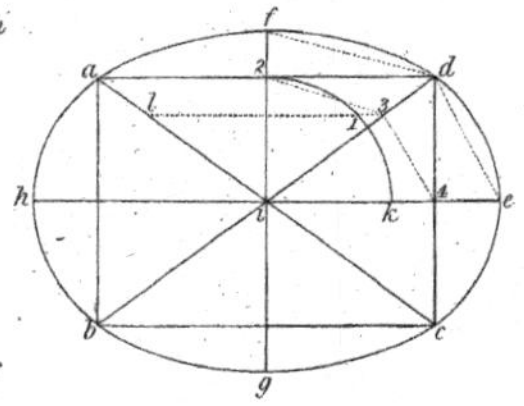

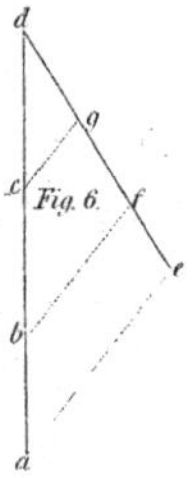

Fig. 6.

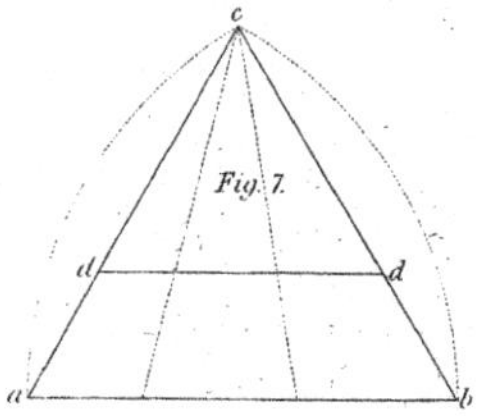

Fig. 7.

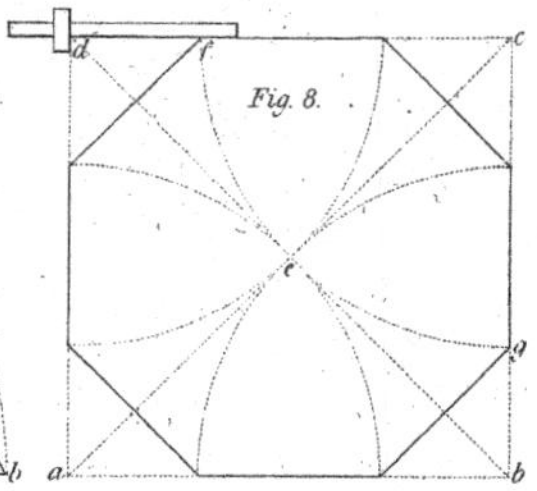

Fig. 8.

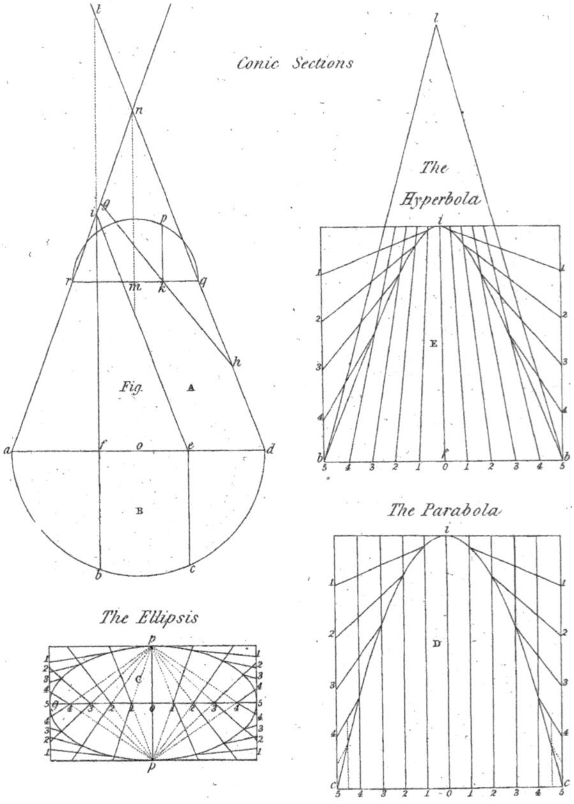
Conic Sections
Fig. A
B
The Hyperbola
E
The Parabola
D
The Ellipsis
C

PLATE 7.

CONIC SECTIONS BY INTERSECTING LINES.

DEFINITIONS.

1. *A cone is a solid having a circular base, from which the sides continually diminish in straight lines to a point in which they all terminate, and this point is called a vertex.*
2. *Opposite cone is another cone joining the vertex of the other cone, with its sides everywhere in the same straight line passing through the vertex as a common point.*
3. *A right line joining the vertex and the centre of the base is called the axis.*
4. *If a cone be cut by a plane passing entirely through its curved surface, but not parallel to the base nor to the axis, nor to a plane touching the side of the cone, the section is an ellipsis, excepting in one position which is a circle.*
5. *If a cone be cut by a plane parallel to its sides, the section is a parabola.*
6. *If the cone be cut by a plane passing through the opposite cone, the figure will be a hyperbola.*

To describe the Ellipsis from the Cone.

FIGURE A. Let B be half the circle of the base of the cone, n the vertex at the top; then $n\ a$ and $n\ d$ are two sides; let the cone be cut by a plane passing through $g\ h$; bisect $g\ h$ at the point k, and through k draw $r\ q$, parallel to the base $a\ d$; also, bisect $r\ q$ in m, describe the semicircle $r\ p\ q$, draw $k\ p$ at right angles to $q\ r$; and $g\ h$ is the length of the ellipsis, and $h\ k$ half its width; from which the figure may be described at C, as explained in the next plate.

To describe the Parabola from the Cone.

FIGURE A. Let $i\ e$ be the axis of the parabola, parallel to the other side $n\ d$ of the cone, and through e draw $e\ c$ at right angles to the base; then will $e\ c$ be half the width of the parabola, and $e\ i$ its height; then the *figure* will be described, as at D, by intersecting lines upon each ordinate, up to the crown, from the equal divisions on each side.

To describe the Hyperbola from the Cone.

FIGURE A. Let the axis of the hyperbola be $i\ f$, cut by a plane passing through f and i, till it cut the opposite cone at l; draw $f\ b$ at right angles to $a\ d$, then is $f\ i$ the height of a hyperbola, and $f\ b$ half the width of the base, and $i\ l$ its transverse axis; then make $f\ i$ at E equal to $f\ i$ in *figure A*, make $i\ l$ in E equal to $i\ l$ in *figure A*, $b\ b$ in E equal to twice $f\ b$ in *figure A*; let the base $b\ b$ in E be divided into ten equal parts, as at 0 1 2 3 4 5, *that is*, into five equal parts on each side from the centre, and draw lines to the point l through these points; likewise divide the height into five each way, and draw lines to the vertex at i; this will show the points through which the curve must pass.

3

PLATE 8.

How to draw any Semi-ellipsis upon the transverse or conjugate Axis, or even a Semicircle itself, by a new Method of intersecting Lines.

FIGURES A and B. Let the given axis be *a b*, and let it be divided into any number of parts, as 10; also let the height be divided into half the number of parts; make *e d* equal *e c*, *that is*, to the height of the arch; then, from the point *d*, draw lines through the equal divisions of the axis *a b*; likewise, through the points 1, 2, 3, 4, 5, in the height *a f*; draw lines tending to the vertex at *c*, which will intersect at the points *h*, *i*, *k*, *l*; and lines being drawn through the divisions of *b g* to *c*, at the crown in the same manner, will give the points *n*, *o*, *p*, *q*; a curve being traced through these points, will form the true curve of an ellipsis.

The semicircle, *figure* C, is drawn in the same manner, by making *a f* equal to one half of *a b*.

How to draw the true Segment of a Circle, by the Method of intersecting Lines.

FIGURE D. Let *a b* be the length of the segment, and *o c* its height, and draw the chord *b c* for one half of the segment, and draw *b m* at right angles to *b c*; and from the point *o* divide *a b* each way, into five equal parts; also from *c*, divide *c m*, and *c n*, each into five equal parts; and draw 1 1, 2 2, 3 3, 4 4, 5 5, on each side, through the divisions 1, 2, 3, 4, 5, on *a s*, and 1, 2, 3, 4, 5, on *b r*; draw lines to *c*, which will intersect the other lines at the points *d*, *e*, *f*, *g*, and *h*, *i*, *k*, *l*: the curve being traced, the thing is done.

How to draw a flat Segment of a Circle nearly true.

Divide the length of the segment into equal parts each way, from the centre *d*, as before, and draw the lines 1 1, 2 2, 3 3, 4 4, 5 5, all at right angles, to the length *a b*; lines being drawn to the crown at *c*, from the divisions at each end, will show the points which the segment must pass through; the curve being traced, the thing is done.

Remark. Although this last method is not the true segment of a circle, but a parabolic curve, yet it will be found useful in practice, in tracing any segment whose height is not more than one-tenth part of its length; if the centre of the segment is found, and drawn with a compass, the difference will hardly be visible, and the flatter the segment, this difference will become the more imperceptible; but if the height exceeds one-tenth of its length, the difference will be visible; for then the arch will be quicker at the vertex, and get flatter and flatter towards each extreme.

In the same manner may all kinds of rampant ellipses be described, or any segment of them, as at *F* and *G*, also a rampant parabola in the same manner, as *H*.

Plate 8.

Fig. A.

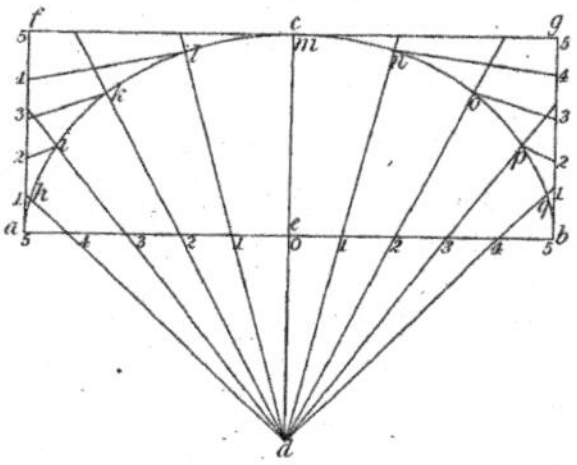

Fig. B.

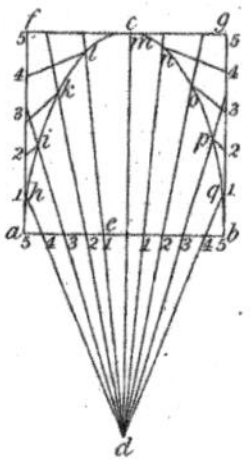

Fig. C.

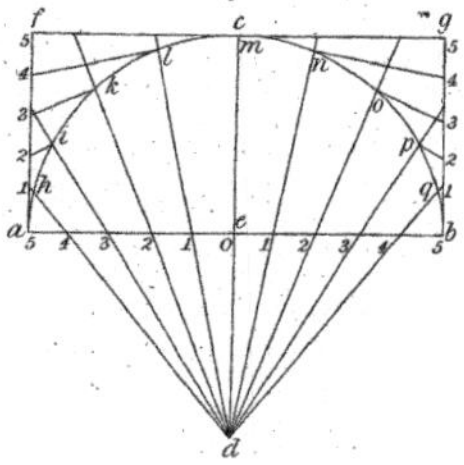

Fig. D.

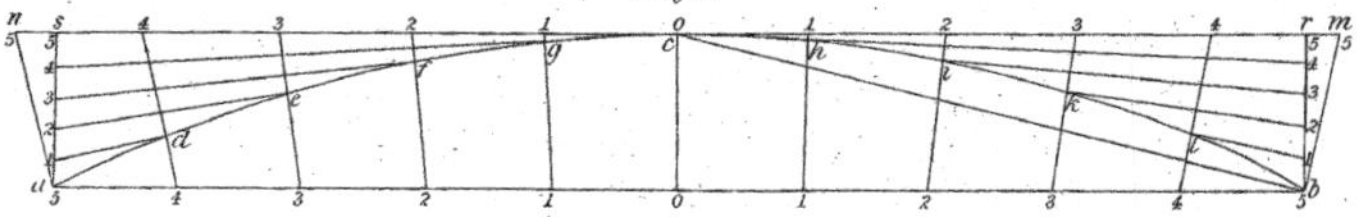

Fig. E.

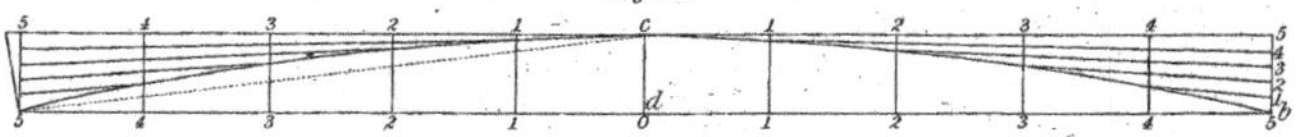

Fig. F.

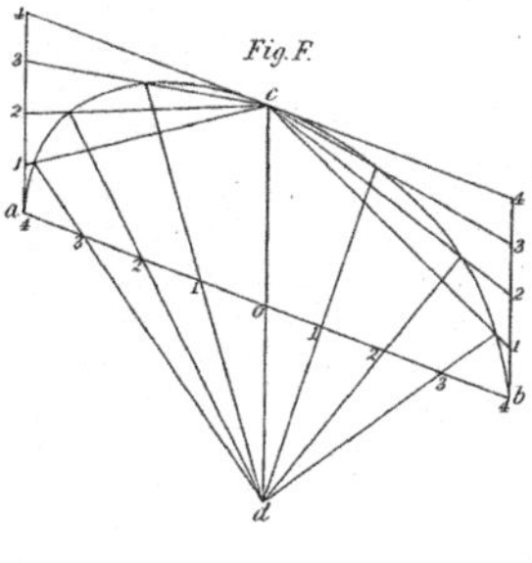

Fig. G.

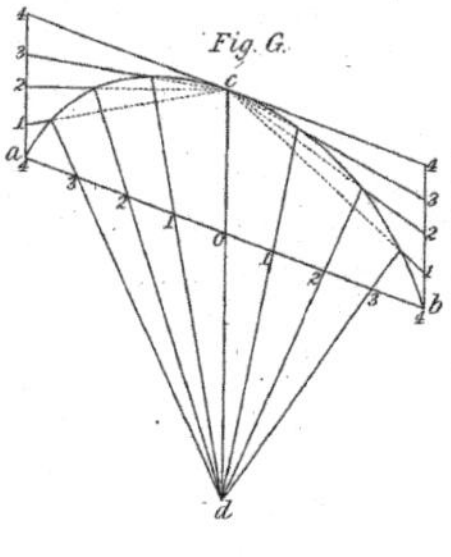

Fig. H.

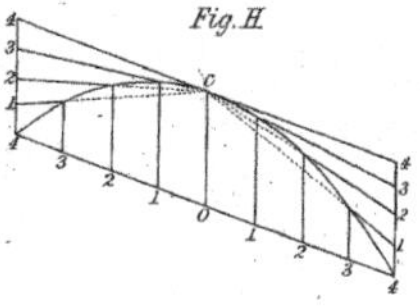

Plate 9.

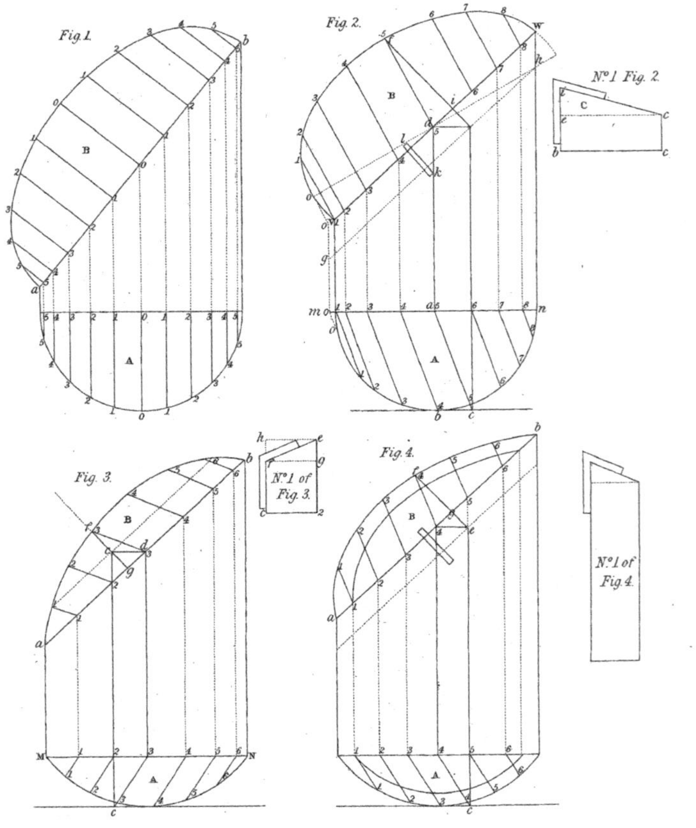

PLATE 9.

THE SECTIONS OF A CYLINDER.

DEFINITIONS.

A cylinder is a figure generated by the revolution of a right-angled parallelogram about one of its sides; consequently the ends of the cylinder are equal circles, and the line passing through the centre of the cylinder is called the axis.

The section of a cylinder, cut by any plane, is an ellipsis, or circle, or rectangle, proved by the writers on Conic Sections.

To find the Section of a Semicylinder, by Ordinates, when it is cut at right Angles to the Plane passing through its Axis, in the Direction a b. Fig. 1.

Let the circle of the base be divided into equal parts at *A*, and drawn parallel up the cylinder to the line *a b*, at the points 0, 1, 2, 3, 4, 5, &c., and from these points draw lines at right angles to *a b*; then *B* being pricked from *A*, as the figures direct, *B* will be the section of the cylinder.

DEMONSTRATION.

Conceive the semicircle *A* at the base to be turned at right angles to the plane, also the semi-ellipsis *B* at right angles to the same plane; then will the ordinates of *B* be parallel and perpendicular over the ordinates of *A*, and every corresponding point in the circumference of *B* will fall perpendicular to the same corresponding points in *A*: therefore *B* is the true section of the cylinder, cut in this position.

To cut a Cylinder in the Direction a b, *upon a Plane, passing through its Axis, to make an acute Angle with the Plane.* Fig. 2.

Let *C*, at No. 1, be the given angle, which the section at *B* is to make with the plane of the cylinder; take *a b* in *figure* 2, *that is*, the radius of the base, and set it from *b c*, at No. 1, perpendicular to *i b*; draw *c c* parallel to *i b*, also from *c* draw *c e* perpendicular to *i b*; then take the distance *c i*, set it from *i* to *f*, in *figure* 2, at *B*; likewise take *i e* from No. 1, and set it from *i* to *e* in *figure* 2, at *B*; draw *e d* parallel to *m n*, to cut the rake in *d*, and join *d f*; then is *d f* the bevel of the first ordinate of the section *B*. And draw the lines *e c* and *d a* parallel to the axis; join *a c* at *A*; then will *a c* be the bevel of the first ordinate of the base. Draw all the other ordinates of *A* parallel to *a c*, and at the points 1, 2, 3, 4, &c. in *m n*, draw lines parallel to the axis of the cylinder, to cut the raking line *V W* at 1, 2, 3, 4, 5, &c. From these points let lines be drawn parallel to *d f*; then the ordinates of *B*, being pricked from the same corresponding ordinates of the base at A, will give the section of the cylinder.

Note. The point *f* will fall beyond the sweep at the section B.

DEMONSTRATION.

Let the plane *B* be conceived to be turned round the line *V W* to make an angle at the point *i*, with the plane *n m V W* equal to the angle *e i c*, No. 1; and conceive a straight line drawn from *e* perpendicular to the plane *n m V W*, the line thus supposed to be drawn will be parallel to the plane *A* of the base, and the triangle formed by *i f*, *i c*, and the perpendicular from *e*, will be equal and similar to the triangle *c i e*, No. 1: then because *d e* and the perpendicular are both parallel to the base, the line that joins the points *e* and *f* will also be parallel to the base; and because *e d* is equal to *a* 6, the triangle *e d f* will be equal and similar to the triangle 6 *a c* in the plane of the base *A*: and because 6 *c* and the perpendicular drawn from *e* are both in a plane parallel to the axis, the plane passing through *d f* and *a c* will also be parallel to the axis: but *d f* is also in the plane of the section, for the point *d* is the intersection *V W*, and the point *f* will be in the perpendicular drawn from *e*; therefore, if a series of planes be conceived to be drawn through the ordinates of the base parallel to the plane passing through *a c* and *d f*, the intersection of these planes on the plane of section *B* will be parallel to the ordinates of *B*, and every two corresponding lines will be in a plane parallel to the axis, and therefore as the lines formed by the intersections of the series of planes in the section *B*, are equal to those in the base *A*, the extremities are in the section of the cylinder.

To cut a Segment of a Cylinder, in the Direction a b, *to make an obtuse Angle with the Plane of the Segment.*
Fig. 3.

Let No 1 be the angle given, which the section *B* is to make with the plane of the segment; from *f* in No. 1, draw *f g* at right angles to *f c*, and *g e* also perpendicular, to make the right-angled triangle *e g f*. And in *figure* 3, at *B*, draw *g f*, at right angles to *a b*, and make *g e* equal to *g e* at No. 1. Also, make *g f* at *B* equal to *g f* at No. 1. Draw *e d* at *B*, parallel to *M N*, and at the point *d*, where it intersects the line *a b*, join *d f*; then *d f* is one of the ordinates. From *e* and *d*, draw the two parallel lines *e c* and *d* 3, join *c* 3; then *c* 3 will also be an ordinate of the base. Draw parallel lines at discretion to *c* 3, for the other ordinates of the base; and from their intersection upon *m n* draw lines parallel upon the cylinder, to cut *a b* in 1, 2, 3, 4, &c., and from these points draw parallel lines to *d f*, which are the ordinates of *B*; these, being pricked from the base as the figures direct, will give the points through which the curve must pass, which being traced, will be the true section of the segment of the cylinder.

DEMONSTRATION

Is the same as the preceding Demonstration.

That the reader may perceive this more clearly, the best way is to draw those lines on pasteboard. The section and the end being made to turn round, in their proper position, then the demonstration will be clearly seen.

Figure 4 is to be laid down and demonstrated in the same manner as Figure 2.

Remark. Upon these figures depend the whole principles of hand-rails for stairs. The reader ought to understand how to form the section of a cylinder, in any case whatever; for the face of raking mould of a hand-rail is nothing but the double section of a cylinder, as in *figure* 4, at *B*, where the double circle upon the base *A* represents the plan of a rail, and the bevel at No. 1, *figure* 4, represents the spring of the plank, and *a b* the pitch of the rail: therefore, it is very necessary that the reader should have a knowledge of these figures and their demonstrations; and not be satisfied with only doing it, but read these demonstrations, and consider them with attention, then he will be able to see the reason why every line is drawn in the manner it is.

Plate 10.

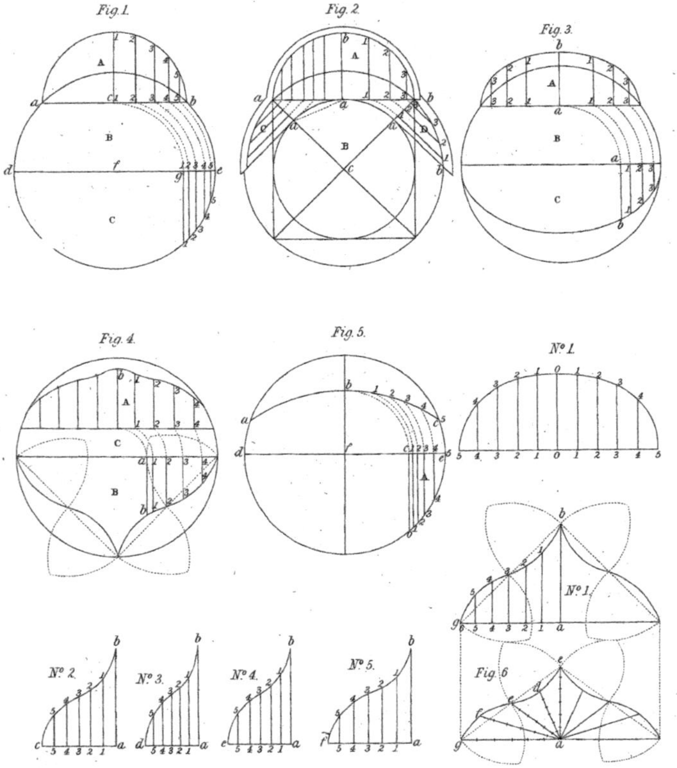

PLATE 10.

THE SECTIONS OF A GLOBE, OR ANY OTHER FIGURE STANDING UPON A CIRCULAR BASE: ALSO, THE SECTION OF ANY FIGURE STANDING ON AN IRREGULAR BASE.

DEFINITION.

A globe is a figure generated by the revolution of a semicircle round its diameter, which becomes the axis of the globe.

AXIOMS: OR, SELF-EVIDENT TRUTHS.

1st. *From this definition it appears, that every two sections passing through the centre, are equal to each other.*

2d. *Every section of a globe, cut by a plane, is a circle; for the generating circle may be made to revolve round any line, as an axis; and therefore every point in it will generate a circle, whose diameter must be twice the radius of that circle distant from the axis of the globe.*

3d. *If a semi-globe is cut by a plane at right angles to the plane of its base, the section will be a semicircle.*

To find the Section of a Semi-globe cut by a Plane at right Angles to the Plane of its Base. Fig. 1.

It appears from the last axiom, that there is no tracing required: for, let the section be cut across *a b*, *figure* 1; divide *a b* in two equal parts at the point *c*; and on *c*, as a centre with the radius *c a* or *c b*, describe the semicircle *A*, which is the true section required.

The same Ordinates. Fig. 2.

Draw any line *d e* through its centre, and let *a b* be the place of the section upon the base, as before; place the foot of your compass in the centre of the globe at *f*, and, with the radius *t c*, draw an arch from *c* to *g*, in the diameter *d e*; the foot of your compass remaining still in *f*, draw the concentric dotted circles from *c b* to *f e*, and at the intersecting points 1 2 3 4 5 in *f c*, and likewise in *c b*, erect perpendiculars to those lines; then *A* being pricked from *C*, as the figures direct, will give the points through which this semicircle must pass.

DEMONSTRATION.

Conceive the semicircle *C* to stand at right angles upon *d e*, also the section *A* to be at right angles to *a b*; now it is evident if *g i* is the height of the globe over the point *g* in the base, *c* 1, which is equal to *g* 1, must also be the height of the section, because the points *c* and *g* stand at an equal distance from the centre; and therefore the point 1 over *c*, is in the surface of the globe. In the same manner it may be proved, that any other points carried round by the dotted lines are in the same surface; but the section that stands upon *a b*, in *A*, is a semicircle; and consequently the method of tracing is also a semicircle.

Observation. Hence appears the erroneous principle of tracing used by a late writer upon this subject, as you may see at *figure* 2, where *A* is the section of a globe, and the bracket at *D* is the section across the diameter. *A* is truly traced from *D*, because the ordinates are carried round in circles; but by his method of tracing, as you see at *C*, upon the other side, the point of the bracket *C* falls within the sweep of the circle, by reason of the ordinate of *C* being carried straight through between the two bases, which I have proved to be false. And this he has applied in bracketing up the angles in the square well hole of a stair case, to the circular curb of a sky-light, which if truly done, is nothing else than upon the same principle as the sections of a globe.

Figure 3 is done upon the same principle as *figure* 1. *A* is the section traced from *C*, and wants no other demonstration than what has been given in *figure* 1.

Figure 4 is an ogee section, standing upon a circular base across the diameter; and *A* is the section traced from it, upon the same principles as *figure* 1.

From these examples it is clear that this method of tracing does not depend on the form of the top, but entirely upon the base. These figures are supposed to be generated round an axis; and, as every circle is carried round at an equal distance from the axis, the perpendicular height of the figure, upon any circle, must be the same height in every point throughout the circle: which proves itself to be the only method for anything of this kind.

A Semi-globe being cut by a cylindrical Surface perpendicular to the Plane of its Base, to find the Form of a Veneer that will bend round it. Fig. 5.

Let *d e* be drawn through the centre *f*; and place the foot of your compass in *f*, the centre; and draw the points *b*, 1, 2, 3, 4, which are equally divided from the centre at *b*, in the circular surface, draw the concentric dotted lines round to the diameter *d e*, at 0, 1, 2, 3, 4, and at these points raise the perpendiculars 0 0, 1 1, 2 2, 3 3, 4 4. Take the stretch-out round *b* 1 2 3 4 5, which is one-half; and lay it upon the base of No. 1, each way, from 0 1 2 3 4, &c., and No. 1 being pricked from *A*, *figure* 5, as the figures direct, will give the points through which the curve must pass for the veneer.

DEMONSTRATION.

For, since the section standing upon *d e* is a semicircle, which is equal to the semicircle upon the base; and as the points 1 2 3 4 in the circular surface, stand at the same distance from the centre *f*, as 0, 1, 2, 3, 4, in *d e*; now if the point *o* at No. 1, is made to coincide with the point *b* in *figure* 5, then the height *o o*, standing over the point *b*, will be equal to the height *o o* at *A*; but these points are at an equal distance from the centre, therefore the top of each ordinate will be in the surface of the globe. In the same manner every other point may be proved, when bent round and elevated, to be of the same height, and at an equal distance from the centre with those of *A*; and therefore No. 1 is the true form of the veneer.

To find the Ribs of a Gothic Niche, being the Plan, and No. 1, *the Front Elevation.* Fig. 6.

Take the length of each base upon the plan, and make them the bases of No. 2, No. 3, No. 4, and No. 5; divide each base into five equal parts; also divide the half of No. 1, into six parts, and draw the ordinates from the equal divisions, perpendicular to each base; then prick each from No. 1, as the figures direct, will give the form of each rib. This wants no demonstration.

Plate II.

Lining for a Parallel soffit cutting oblique
In a straight wall

Fig. A.

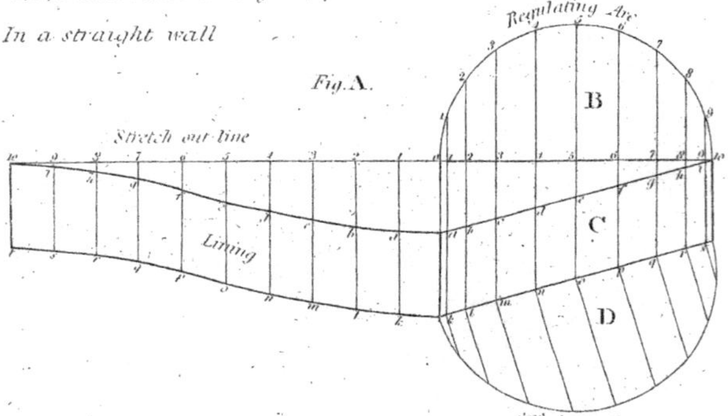

Lining for a Parallel soffit cutting
Right into a Circular wall

Fig. E.

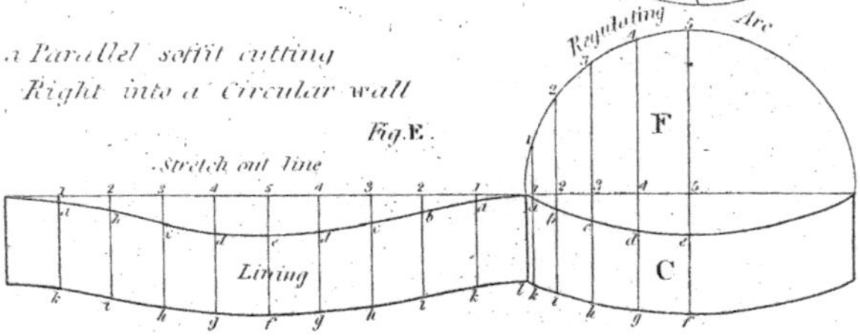

OF CARPENTRY.

LININGS FOR SOFFITS.

DEFINITION.

The lining of a Soffit, in the Theory of Carpentry, signifies the covering of any concave Surface of a Solid spread out on a Plane if possible.

Soffit in Architecture is the under side of the head of a door, window, or the intrados of an arch, and may be either plane or curved.

PLATE 11.

How to stretch out a Soffit, when a Window or Door, having a semicircular Head, cuts into a straight Wall, in an oblique Direction.

LET *C* be the plan or opening of the window, in *fig. A*, and let the base of the semicircle *B* be drawn at right angles to the jambs, or sides of the plan *C*; divide the semicircle into any number of equal parts, as ten, and draw the ordinates across the plan, extend the parts round *B* upon the stretch-out line, the ordinates being drawn from the divisions across, and traced off from the plan *C*, as the figures and letters direct, the lining of the soffit will then be completed.

If you would make a cylinder to be only the thickness of the wall, *D* shows the end of it, which is to be traced from the semicircle *B*.

How to draw the lining of a Soffit when the Top is a Semicircle, cutting right into a circular Wall.

FIG. *E*. This and the other below are performed the same as that above, with this difference, that you are to prick from the circular plan, instead of the straight plan.

FIG. 1 shows the method when a circular headed window cuts oblique into a circular wall.

Note. In all kinds of cylindro-cylindric soffits, when the two jambs are parallel, the straight line, which the soffit is pricked from, must be drawn at right angles to the jambs, as is shown in this plate; for want of this consideration, they are shown in books upon wrong principles.

But in the following soffits, where the jambs are not parallel, they must be continued till they meet in a point, and the line which the soffit is to be pricked from, must be made to form an isosceles triangle with the jambs.

PLATE 12.

To draw the lining of a Soffit in a straight Wall, splaying equally all round with a circular Head.

In *fig. A*, continue the sides of the plan *A*, that is *a c* and *b d*, to meet at *e*; then about the centre *e*, and from the points *a* and *c*, describe the soffit *C*, and stretch the semicircle *B* along the outline of the soffit *C*, it will be completed.

To draw the lining of a Soffit in a circular Wall, splaying equally all round with a circular Head.

FIG. *B*. The stretch-out of this soffit is managed the same as in the last; draw the ordinates of the semicircle *B*, from thence continue them to *f*, the concourse of the splay, and at the points *a*, *b*, *c*, *d*, *e*, where they intersect the plan, draw the parallel lines *a e*, *b f*, *c g*, &c. parallel to the base of *B*, and from the points *e*, *f*, *g*, *h* and *i*, circle lines to *a*, *b*, *c*, *d* and *e*, round the centre *f*, which will give the half of one edge of the soffit, the other half being pricked from it; the other edge is found in the same manner.

Note. This cannot be pricked from the plan as the others are, as the lines round the splay are not level with the plan, and will therefore be longer than those on the plan.

DEMONSTRATION OF FIG. *A*.

Conceive the semicircle *B* to be turned at right angles to the plan *A*, then every point in the circumference of the semicircle *B* will be at an equal distance from the point *e*, but the soffit *C* is described with the same radius; therefore the edge of the soffit *C*, that is, the arch line *a f* will exactly coincide with the arch of the semicircle *B*, which was to be proved.

DEMONSTRATION OF FIG. *B*.

It is easy to conceive from the last demonstration, that if the semicircle *B* is turned up, and the soffit at *C* bent round it, the points 1, 2, 3, 4, 5, at *C*, will coincide with equal divisions in the semicircle *B*, and the points *a*, *b*, *c*, *d*, &c. at *C*, will fall perpendicularly over the points *a b c d*, &c. in the plan *A*: for the arches *a e*, *b f c g*, *d k*, and *e i* at *C*, will fall over the parallel straight lines *e a*, *f b*, *g c*, *h d*, *i e*, in the plan *A*, which was to be demonstrated.

The learner is advised to cut these and the following soffits out of pasteboard, and their demonstrations will be more clearly seen.

Plate 12.

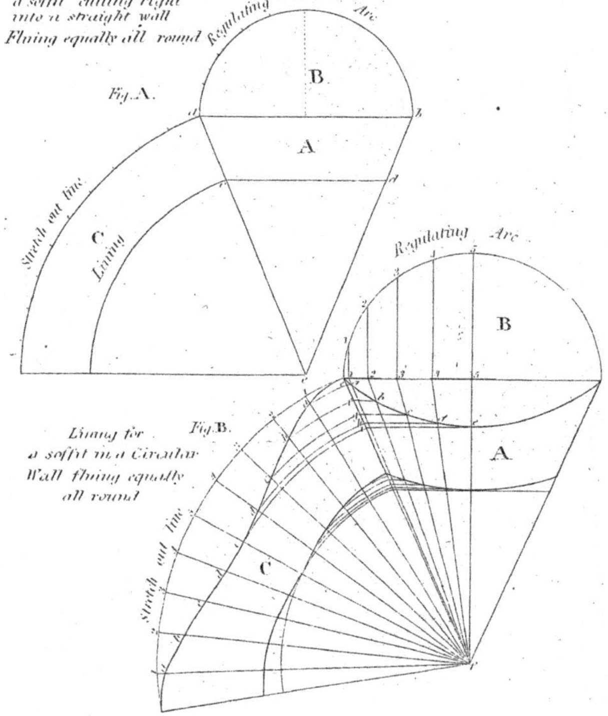
Lining for
a soffit cutting right
into a straight wall
Flning equally all round
Regulating Arc
Fig. A.
B
A
C
Stretch out line
Lining
Regulating Arc
B
A
C
Fig. B.
Lining for
a soffit in a Circular
Wall flning equally
all round
Stretch out line

Plate 13.

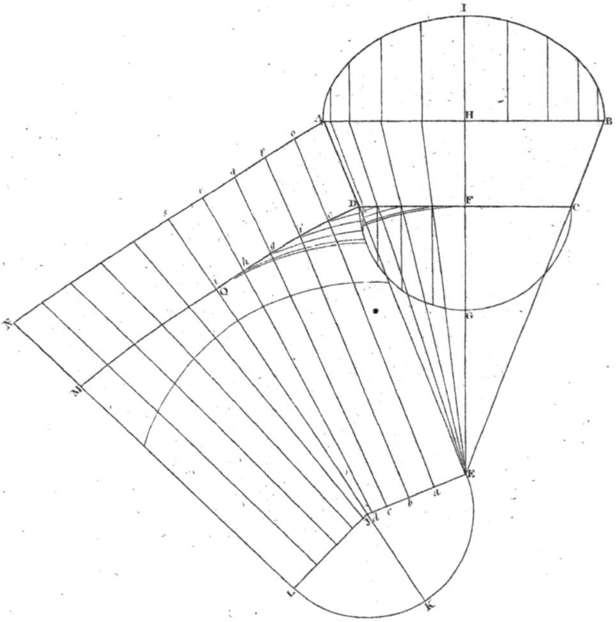

PLATE 13.

To find the lining of an Aperture whose Plan A B C D *is a Trapezoid, with two parallel Sides* A B *and* D C, *which represent the out and insides of the Wall, and two equally inclined Sides* A D *and* B C, *which represent the Jambs, and whose elevation* A I B, *on the inside of the Wall is a semi-ellipsis, and that on the outside* D G C *a Semicircle, so that a straight Edge may everywhere coincide with the lined Surface, and be parallel to the Horizon.*

Produce *A D* and *B C* to *E*; bisect *D C* at *F*, and draw *E F*; produce it to *I* and it will cut *A B* at *H*; then *F G* and *H I* are equal to each other. Divide the quadrant *D G* into any number of equal parts, (as five,) and draw lines through the points of divisions cutting the base *D C*; from the points of division and in the same straight line with the point *E*, draw lines to cut *A H*, and the lines so intercepted will represent the level straight lines on the soffit. Make *E J* perpendicular to *E D* equal to *F G*, and mark the other divisions on *E J* from *E*, at *a*, *b*, *c*, *d*, respectively equal to those in *F D*: then take the distances of the several points in *D C* from *E*, beginning next *E D*, and proceeding to the last *E F*, and describe the arcs from the centres, *a*, *b*, *c*, *d*, respectively; with the fifth part of the arc *D G* fix the foot of the compass in *D*, and cross the first arc at *e*; place one foot of the compass in *e*, cross the next arc at *f*, proceed in this manner to *i*, then *D*, *e*, *f*, *g*, *h*, *i*, will be the coincident line of the lining or interior covering for the arc *D G*. Join *i J* and produce it to *K*; make the angle *K J L* equal to the angle *K J E*, and make *J L* equal to *J E*: mark the divisions on *J L*, so that the distances from *J* may be equal to the distances of the several divisions on *J E*, then the other half of plano-cunioidal line may be found by inversion. Produce the lines *a e*, *b f c g*, *d h*, *j i*, &c., to *o*, *p*, *q*, *r*, *s*, &c., make *e o*, *f p*, *g q*, &c., in inverted order equal to the seats of the lines on the soffits and the points *o*, *p*, *q*, *r*, &c., then curves being drawn through the points *o*, *p*, *q*, *r*, *s*, &c., and through *e*, *f*, *g*, *h*, &c., will form the wall lines of the covering, so that *A D M N* will be the whole covering or interior development.

PLATE 14.

To find the lining of an Aperture covered with the same Surface, and terminated by a circular Wall.

Find the line *D Q M*, as in the last plate, as in a straight wall, then transfer the distances from the straight line on the plan to the arcs, representing the faces of the wall to the covering, and the edges will be obtained as in the last.

Plate 14.

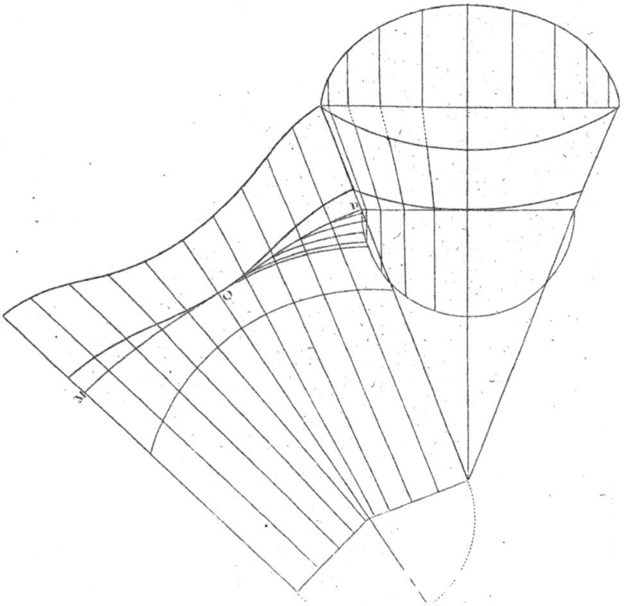

Plate 15.

Fig. A.

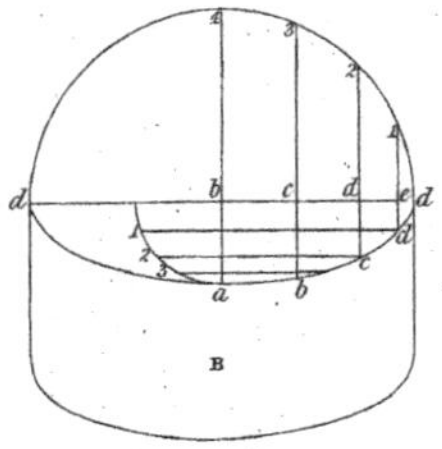

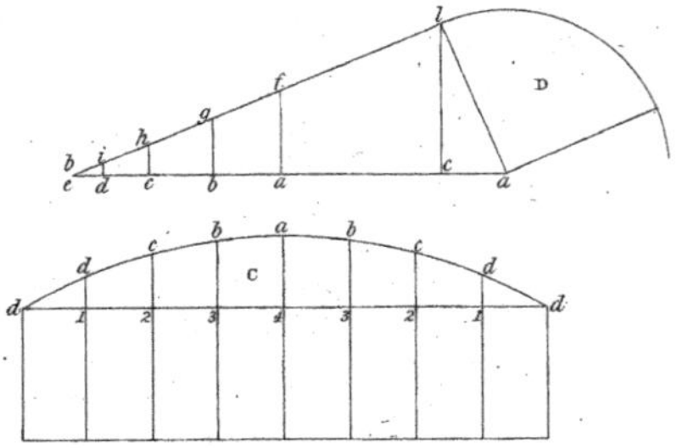

Fig. E.

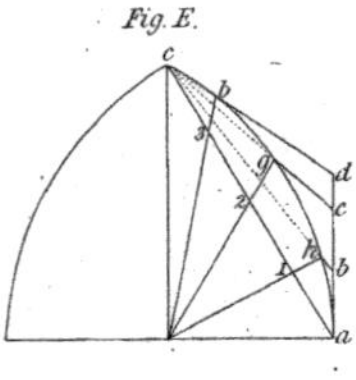

N° 2.

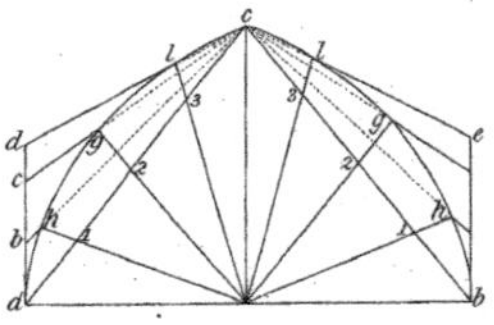

N° 3

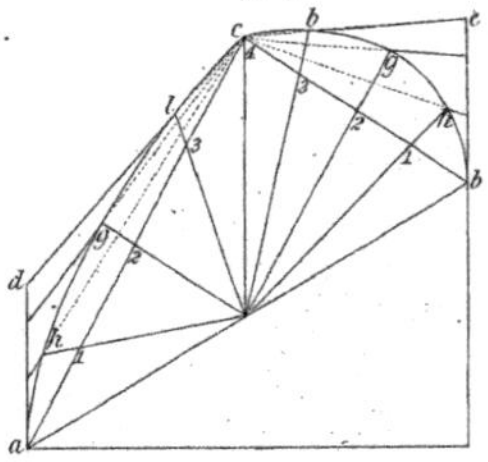

Fig. F.

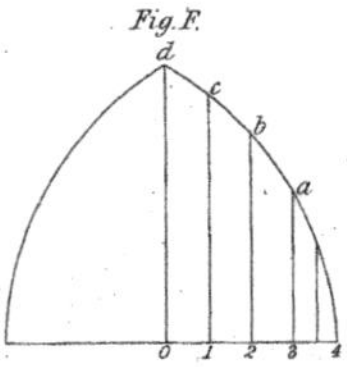

Fig. F. N° 1.

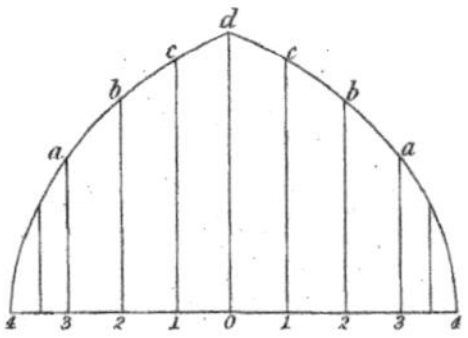

Fig. F. N° 2.

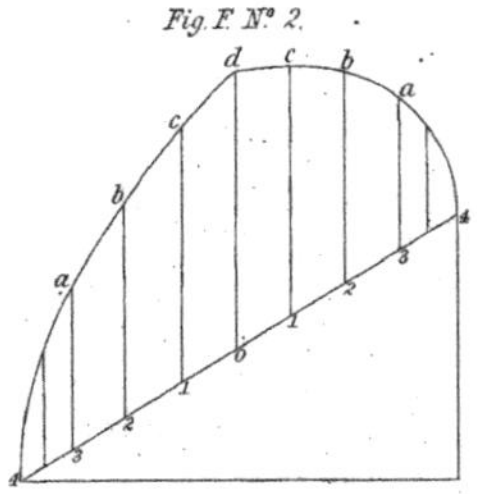

PLATE 15.

To draw the lining of a cylindrical Soffit, cutting right in a Wall which does not stand perpendicular to the ground to a level Base. Fig. A.

Let *a e* at *D* be the level of the ground, *a l* the inclination of the wall, equal to the radius of the cylinder; let fall the perpendicular from *l* to *c*, in the bottom line *a e* make the semicircle in *fig. A*; to the width of the cylinder, or the double of *a l* at *D*, take the distance *a c* at *D*, and make *b a* equal to it in *fig. A*, and describe a semi-ellipsis to the length of the semicircle *d d* and to the width *a b*; lay the equal divisions round the semicircle in *fig. A* in *C*, along the line *d d*, on each side of the middle point 4, then take the parts *e d*, *d c*, *c b*, *b a*, from the plan *B*, and lay them at *D* respectively from *e* towards *d*, *c*, *b*, and from *l* draw *l e* to make a right angle with *l a*, and at the points *a*, *b*, *c*, *d* erect perpendiculars to *a e* to cut *l e* at *f*, *g*, *h* and *i*: take the distances *e i*, *i h*, *h g*, and *g f*, in *D*, and lay them on the soffit at *C* respectively, from 1 *d*, 2 *c*, 3 *b*, 4 *a*, each way, then will the straight line *d d* in the soffit, when bent round, be perpendicular over the elliptic line in the plan *B*, and the curve line *d d c b a*, &c., *d* will fall over the points *d d c b a* in the plan: in the same manner the edge of the soffit may be brought to answer any curve line proposed.

To draw the Arches of Groins by a new Method, whether right or rampant, so that their Arches shall intersect or mitre truly together, from a given Arch of any Form.

Let *fig. E* be the given arch of a Gothic form, draw the chord *a c* for one-half the arch, divide it into any number of parts, as four, and through the divisions draw lines from the centre *e* to terminate in the circumference at *h*, *g*, *l*, draw lines from *c* through *h g l* to cut the perpendicular *a d* at *b*, *c*, *d*; and if No. 2 is required to be wider, but the same height as *fig. E*, draw the two chords *a c* and *c b* for each side of the arch, divide each into four equal parts, as before, and set the parts *a b*, *b c*, *c d* perpendicular on each end of *a b* at No. 2, and from the divisions draw lines to the vertex at *c*, then trace the curve through the points *h*, *g*, *l*, &c., so the arch at No. 2 will truly mitre into *fig. E*; in the same manner the rampant curve at No. 3 will be brought to correspond with *fig. E*, and No. 2.*

* It is hardly possible to find a more ready method in practice, because a chalk line will soon strike all the radical lines, having only to move it but once from the point *e* up to *c* at the crown; *fig. F.* shows the common method by dividing the basis of each into a like number of parts, and transferring the height, as the figures explain, at No. 1 and No. 2; nothing is more tedious in practice than raising a number of large perpendiculars, and going continually from one curve to get the height of another.

PLATE 16.

As it happens sometimes in church work, that windows go higher than the ceiling line, which therefore requires to be hollowed out, so that the light may be thrown down into the body of the church: I shall in this place show the method of making a curb for that purpose.

To Find the Form of the Curb.

Let *k b l* be the head of the window, *figure A*, and let it come as high as *a b*, above the ceiling;* and let *a b* at No. 1, be the same height, and *b c* the direction of the light, and *a c* will be the length of the curb. Make *a c* at No. 2, equal to *a c* at No. 1, and divide it into six equal parts; also divide *a b*, in *figure A*, into six equal parts, and let the ordinates be drawn as is explained in the figure; a curve being traced round the points of intersection, will give the form of the curb.

Figures *B* and *C* show the method of drawing and backing any elliptic rib with a compass, which is exceedingly handy in drawing, and will be near enough for the representation of an elliptic rib on paper, as no other method will be so clean when done; but for practice, a trammel, or intersecting lines is more ready.

To draw and range the Ribs by this Method.

In *B*, let *c h* be the height, and *c b* the width; divide the difference into three equal parts, and set four such parts on each side of *c*, to *d* and *d*, and make an intersection with the distance *d d* at *e*, and draw a line through *e* and *d* to *i*, then *d* and *e* are the centres for the interior side: suppose the rib is to be backed as much as *a b* upon the bottom, set *a b* from *d* to *f*, and from *e* to *g*, parallel to the base; and draw a line through *g*, *f*, to *k*: then *g* and *f* are the centres for describing the ranging lines.

The rib *E* is traced from *D*, and *a b* being set all round on the parallel lines, shows how the ranging is found for a drawing on paper: the rib *C* is described in the same manner.

The word backing is properly applied to the upper side of anything in Carpentry or Joinery, as the back of a rafter, the back of a rail; but range applies either to the operation of levelling the upper or lower edge, and explains its own meaning, viz. forming the edge so as to range with the other edges, whether forming a ceiling or the exterior of a roof.

* The ceiling is here supposed to be level, which is seldom the case in a church; but the method will be nothing different if the ceiling line *a e* were to incline to the horizon in any angle whatever, only observe to make *c a b* equal to that angle.

Plate 16.

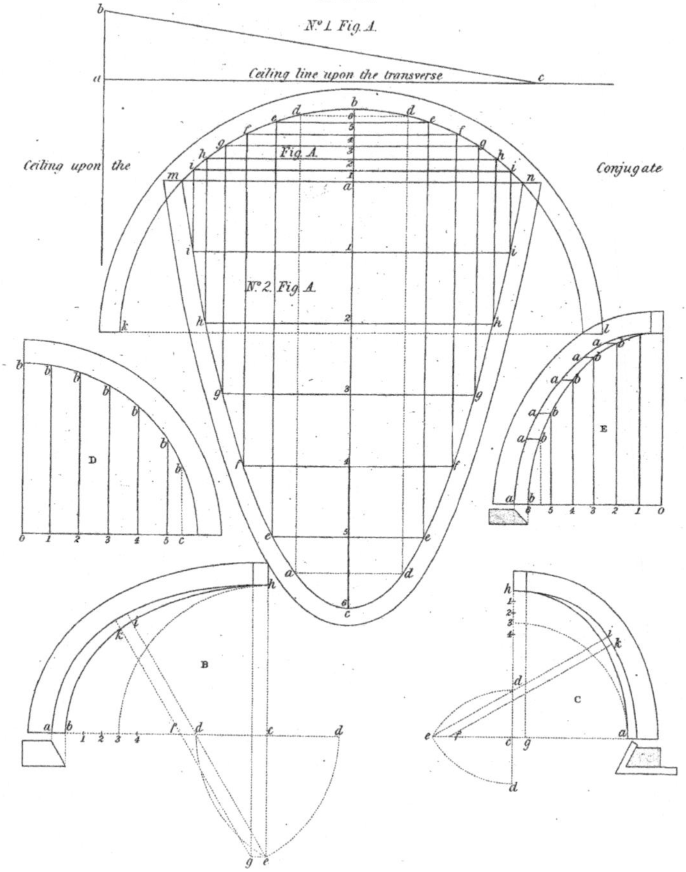
N.° 1. Fig. A.
Ceiling line upon the transverse
Ceiling upon the
Conjugate
Fig. A.
N.° 2. Fig. A.
D
E
B
C

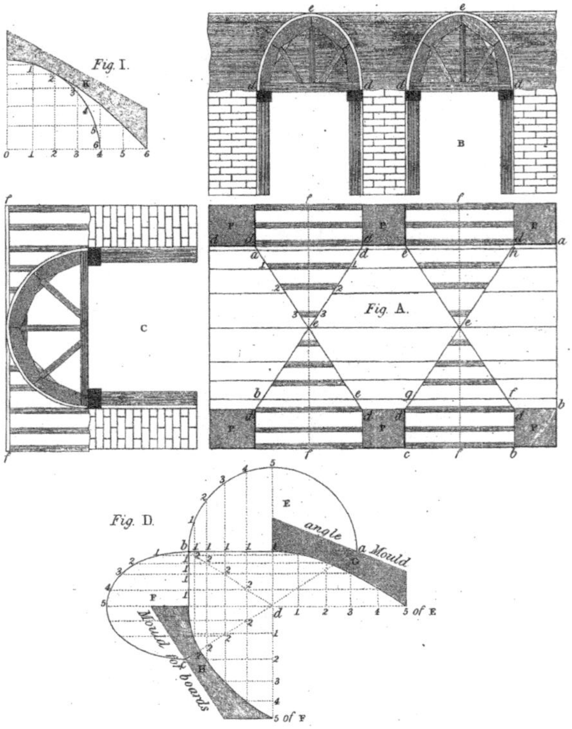
Fig. I.
Fig. A.
Fig. D.
angle
Mould
Mould for boards
B
C
E
F
G
H
5 of E
5 of F

PLATE 17.

DEFINITION.

Groins are formed by the intersection of arches or vaults, and the surfaces of their meeting may be considered as the sections of cylinders, cylindroids, &c.

BRICK GROINS. Description.

P, P, P, &c., is the plan of the piers which the vault is to stand upon, *a b, fig. D,* is the end opening, which is a given semicircle; and *b c* is the opening of the side arch, which is to come to the same height as the end arch *a b*: fix your centres over the body range, *fig. A,* as shown in the section at *C,* then board them over. In *fig. A,* is the manner of fixing the jack ribs upon the boards, which likewise shows at *C.*

To find the Mould for the Jack Ribs.

Take the opening of your arches in *fig. A,* that is, *a b* and *a d,* and lay them down in *fig. D,* at *a b* and *b c,* to make a right angle. Divide one-half of the given semicircle into five parts, and square them across 1, 1, 1, &c., to cut *b d* and *d c,* the diagonals in 2, 2, 2, &c., and through the points 2, 2, 2, &c., draw lines parallel to 1, 1, 1, &c., the base of *E,* both ways towards *F* and *G*; stick in nails at 1, 2, 3, 4, 5, in the arc of *E,* and bend a thin slip of wood round them, which mark with a pencil at every nail; this slip of wood being stretched out from *d,* 1, 2, 3, 4, 5, and squared over to *G,* will intersect the other lines in small rectangles: a curve being traced through the diagonals of each rectangle, will give a mould to set the jack ribs.

How to fix the Jack Ribs.

Bend your mould *G* from *a,* to the crown at *e,* in *fig. A,* that will give the edges of your boards; then fix a temporary piece of wood, level upon the crown, in the direction of *f, f,* and let it come the thickness of your boards lower than the crown, and it will give the height of your jack ribs, which is a very sure method of placing them.

To find a Mould to cut the ends of the Boards.

The rib *F* is traced to the height of *E,* or got by a trammel, which will be fully exemplified in the following plates. Take the parts round *F,* and lay them out to 1, 2, 3, 4, 5; then *H* will be got in the same manner as *G,* which will be a mould to cut the ends of your board that goes upon the jack ribs against the body range.

Fig. 1. *Is an easy method of getting the moulds when both arches are the same opening.*

Take half the opening of the arches, whatever they are, and draw a quarter circle, and divide it into six; bend a slip round it to take its parts, then stretch it out upon the base from 0 to 6, and square over your points 1, 2, 3, &c. Through the points in the arch draw the lines on both sides parallel to 0, 6, the curve being traced as before, gives both moulds of an equal and similar form.

Note. The curve *F* may be drawn in practice with a trammel, independent of the other, and the two moulds *F* and *G* may be drawn separate, without any connection of lines, as shall be shown hereafter.

PLATE 18.

DEFINITIONS.

Groins are said to be ascending or descending when they are not built upon level ground.

CENTRING FOR ASCENDING OR DESCENDING GROINS.

The Plan and Inclination of any Groin being given, and one of the Body Ribs, as B, *also the Place of the Angles upon the Plan, to find the Form of the Side Ribs, so that the Intersection of both Arches will be perpendicularly over the Plan.*

Divide half the circumference of the given rib *B*, into any number of equal parts, and draw them to intersect the angles; and from thence let them be returned up to the rib *C*, upon the side; then *C* being pricked from the given rib at *B*, as the letters direct, will give the form of the side centre. The same is shown at *F*, by the method of intersecting lines.

To find the two moulds D *and* E *for placing the Jack Ribs, to bend over the Angles in the Body Range, when boarded in, so that they may be perpendicular over the Angles upon the Plan.*

At *C*, draw lines from the points *a*, *b*, *c*, *d*, *e*, *f*, *g*, &c., where the ordinates of *C* intersect the top of the arch, perpendicular to the rake, and draw the semi-ellipsis, *A*, to the width of the body range; and to *a h*, the height of the side centres, perpendicular to the rake; and continue the ordinates of *B*, up to *A*, to intersect at 1, 2, 3, 4, 5, 6. Bend a slip round these points, and mark them opposite to every point, and stretch it out along *k*, 1, 2, 3, 4, 5, 6, between *D* and *E*, and draw lines through these points, at right angles to *k* 6, to intersect with the perpendiculars. Begin at 6, and trace a curve both ways, will give the edges of the two moulds for placing the jack ribs.

To cut the Jack Ribs to the Rake of the Groins.

Set the number of the jack ribs upon the arch *B*, at their proper distances, and take their several heights, that is, *h i*, *k l*, *m n*, and set them upon the arch *G*, from *a* to *b*, and from *a* to *c*, and from *a* to *d*, draw lines through these points parallel to the rake, which will show how the jack ribs are to be cut, so that they shall range properly with the other raking centres.

Note. All the body ribs must be ranged according to the rake of the groin; to do this exactly, the under edges of all the ribs must be bevelled according to the rake; then make a mould as *B*, or one of the body ribs themselves will answer instead of a mould, which being applied to each side of any other rib, keeping the bottom fair with the under edge upon each side, and drawing the curves by the other, it will give the ranging line.

Plate 18.

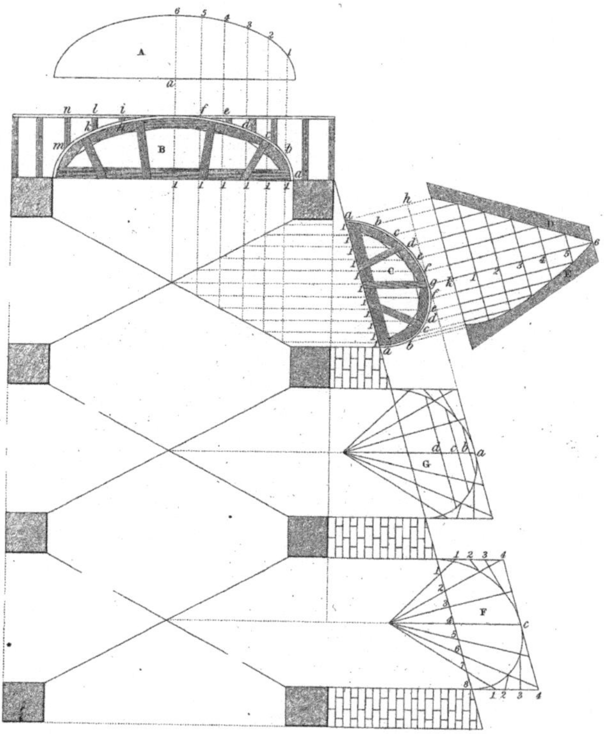

Plate 19

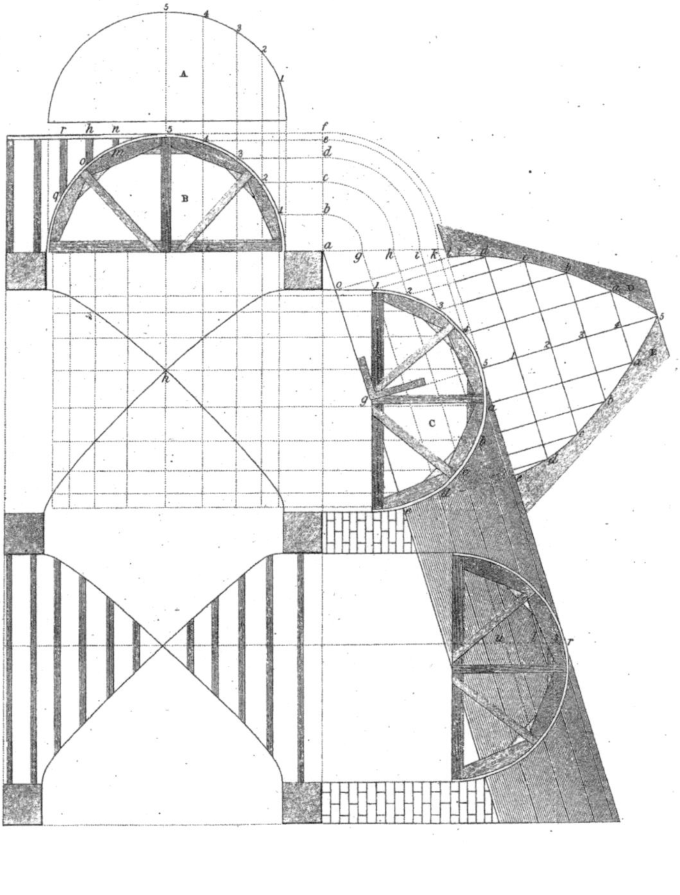

PLATE 19.

Given the two side Arches of any Groin, and the Inclination, to find the Intersection of the Angles upon the Plan.

Divide half of the body rib *B* into equal parts, and draw parallel lines to *b, c, d, e* and *f*; and from the point *a*, as a centre, draw the concentric dotted circles round to *g, h, i, k, l*; then draw parallel lines to the rake, to cut the centre *C* at 1, 2, 3, 4, 5, and *a, b, c, d, e*, on the other side; and from these points let lines be drawn perpendicular through the plan. And on the centre *g* of the rib *C*, square a line up to 5, the top of the arch *C*; and from 5 draw a line perpendicular through the plan. Also through the points 1, 2, 3, 4, 5, at *B*, draw perpendicular lines to the plan the other way; begin at *h*, and trace through the angles both ways, will give the place of the angles upon the plan.

The moulds for bending over the angles are found in the same manner as in the last plate, by taking the stretch-out round *A*, and laying it between *D* and *E*.

The reader may see such groins executed under the Adelphi buildings in the Strand, London, where the declivity is very rapid in going to the river.

The jack ribs of the groin are cut in the same manner as directed in the last plate, and in the practice there will be no occasion for tracing the angles, as the two moulds *D* and *E* are done independent of them; the reader will farther observe, that the arch *B* must not be used instead of the arch *A*, which would produce a very great error in the moulds *D* and *E*, as it must be evident to every one, that the section upon the square of the cylinder, or body range, must be less in the height than the perpendicular or plumb section *B*, which in this case is oblique: if these things are properly understood, there will occur nothing in brick groins but what may be easily surmounted.

In all kinds of brick groins the centres or body ribs must be fixed first in the same manner as if there were no side arches cutting across them; then the centres must be boarded over; then to find the place of the angles upon the boards, that is, the proper intersection of the side arches upon the plan, the moulds *D* and *E* must be both bent round the boards at one time, by keeping the points *l* and *e* of the moulds *D* and *E* upon the tops of the piers at *o* and *e*; then keep the top points together, and bend them round, keeping them still together, then the point at 5 will fall perpendicularly over *h* in the plan; round the inner edges of the moulds draw a curve upon the boards, which will be the proper intersection of the side arch. The jack ribs are cut in the same manner as directed in the last.

PLATE 20.

The Angles or Diagonals of any Plaster Groin, which are straight upon the Plan, and one of the Side Arches being given, to find the other Side Arch and Angle Rib.

FIG. 1. CASE 1. If the given rib is a semicircle or semi-ellipsis, they may be described as in *fig.* 2, plate 6, with a trammel, which is by far the readiest method; but if a proper trammel is not to be got, a temporary one may easily be made, which will answer equally well by fixing two pieces of wood in the form of a square, that is, to make a right angle; each leg must be as long as the difference between the semi-transverse and semi-conjugate axis, and instead of the sliding nuts in the rod, two brad awls will answer the purpose, being put through any straight slip of wood; and by moving this round either the exterior or the interior angles of the square, keeping the pins or brad awls close to each leg, it will describe one quarter of an ellipsis at one time.

To find the length of the Jack Ribs.

Lay down the plan of the ribs as at *B*, and draw a rib upon each opening; then draw perpendicular lines from the plan of each opening, at the extremities *a c e*, to cut its corresponding ribs at *b d f*: then the distance from *b* to *b* shows the length of the first jack rib, from *d* to *d* the length of the second, and from *f* to *f* the third.

How to bevel the Angle Ribs, so that they shall range with each opening of the Groin.

First get the ribs out in two halves or thicknesses, as at *E* and *F*, then draw the plan of your angle rib, which is placed between *E* and *F*, will show the true ranging upon the bottom of the rib; then shift your hip mould parallel upon the base of *E* and *F*, will show how much wood there is to be bevelled off; then nail the two halves together, and it will be completed.

METHOD I.

FIG. 2. CASE 2. When the given rib is a segment of a circle, or any other curve whatever, the ribs will be described as in plate 15, *fig. E*, as are shown at *B*, *E*, and *F*.

METHOD II.

When the given arch is a segment of a circle as at *A*, take its height *b c*, and place it from *b* to *c* at *G* and *D*; then take the whole diameter of the arch *A*, that is, twice the radius *a c*, and place it from the crown of the other arches perpendicular to their bases from *c* to *b* at *C*, and from *c* to *d* at *D*; then the arch may be drawn as in plate 8, by intersecting lines: the ranging of the ribs is done in the same manner as in the last groin.

Either of these two methods is much readier in practice than tracing the ribs through ordinates.

Plate 20.

Fig. 1.

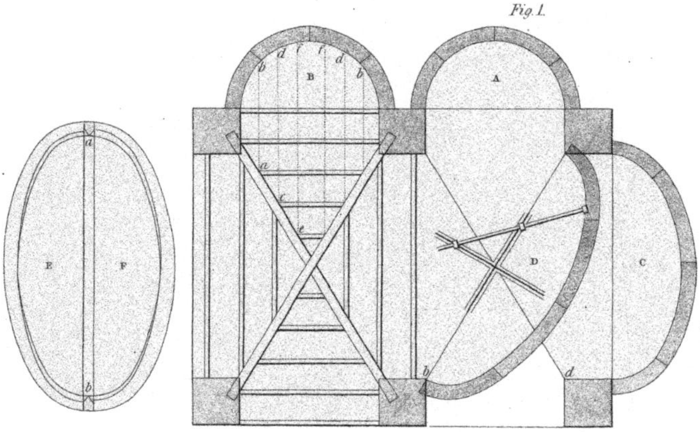

Fig. 2.

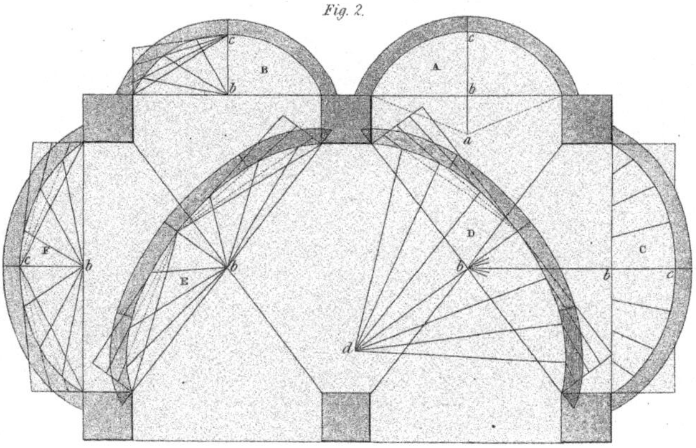

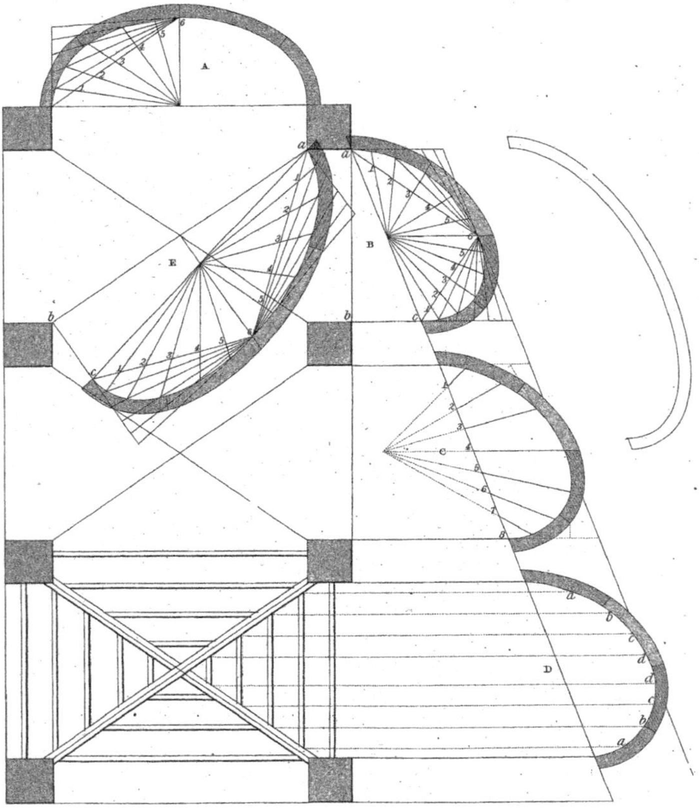
A
B
C
D
E
a
b
c
d

PLATE 21.

Given one of the Body Ribs, and the Angles straight upon the Plan, and the ascent of a Groin not standing upon level Ground, to find the Form of the ascending Arches, and the Angle Ribs.

Let *b a c* at *B* be the angle of the ascent, from the point *b* make *b c* perpendicular to *a b*, and describe the rampant curve *B*, as in plate 15, at No. 3, in *fig. E*: then draw the diagonal *a b* at *E*, and make *b c* perpendicular to it, and equal to *b c* at *B*; then draw the hypothenuse *a c*, and describe the angle rib *E*, in the same manner as that of *B*.

To find the Length of the Jack Ribs, so that they shall fit to the Rake of the Groin.

Draw lines up from the plan to the arch, as at *D*, in the same manner as explained in the last plate; then the arch from *a* to *a* is the first jack rib, from *b* to *b* the second, and from *c* to *c* the third, &c.

How to range the Angle Ribs for such sort of Groins.

Get the ribs out in two halves, as in the last plate, then the bottom of the ribs must be beveled agreeably to the ascent of the groin, and the plan of it must be drawn upon the level, and from thence they may be drawn perpendicular from the plan to the rake of the rib; then take a mould to the form of the rib, or the rib itself, and slide this agreeably to the rake to the distance that is marked upon the bottom to be backed off, will show how much the rib is to bevel all round.

Plate 21.

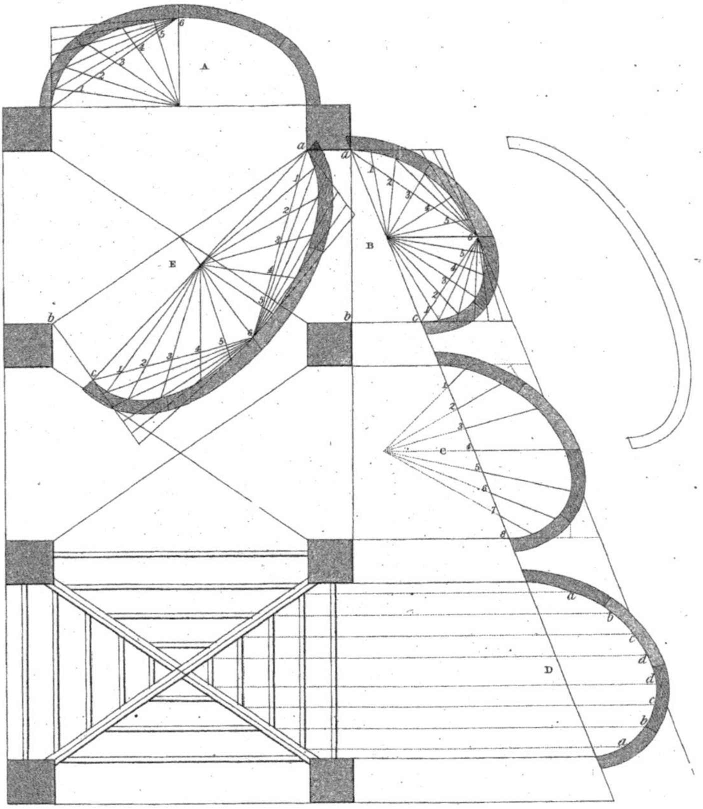

PLATE 21.

Given one of the Body Ribs, and the Angles straight upon the Plan, and the ascent of a Groin not standing upon level Ground, to find the Form of the ascending Arches, and the Angle Ribs.

Let *b a c* at *B* be the angle of the ascent, from the point *b* make *b c* perpendicular to *a b*, and describe the rampant curve *B*, as in plate 15, at No. 3, in *fig. E*: then draw the diagonal *a b* at *E*, and make *b c* perpendicular to it, and equal to *b c* at *B*; then draw the hypothenuse *a c*, and describe the angle rib *E*, in the same manner as that of *B*.

To find the Length of the Jack Ribs, so that they shall fit to the Rake of the Groin.

Draw lines up from the plan to the arch, as at *D*, in the same manner as explained in the last plate; then the arch from *a* to *a* is the first jack rib, from *b* to *b* the second, and from *c* to *c* the third, &c.

How to range the Angle Ribs for such sort of Groins.

Get the ribs out in two halves, as in the last plate, then the bottom of the ribs must be beveled agreeably to the ascent of the groin, and the plan of it must be drawn upon the level, and from thence they may be drawn perpendicular from the plan to the rake of the rib; then take a mould to the form of the rib, or the rib itself, and slide this agreeably to the rake to the distance that is marked upon the bottom to be backed off, will show how much the rib is to bevel all round.

PLATE 22.

OF GROINS CUTTING UNDER PITCH.

DEFINITION.

When the side arches of a groin are lower than the body arch, then they are called under pitch groins.

Given one of the Body Ribs B, *and the height* f g, *of a Door or Window, &c. at* D, *and its Width* m l, *to find the Side and Angle Ribs* D *and* E, *so that the Intersection of the Side Arch* D, *with the Body Rib* B, *shall be straight upon the* Plan.

Draw *c e* perpendicular to *c b*, the base of *B*, and equal to the height of the window at *D*, that is, equal to *f g*; through *e* draw *e a* parallel to *c b*, cutting the arch *B* in *a*; let fall the perpendicular *a b* to *c b*, and continue it so as to cut the line *f g* produced to *k*, and draw *k m* and *k l*, which is the place of the angles upon the plan, or the base of the angle ribs; then the ribs *D* and *E* may be described from the given rib *F*, as directed in plate 15, *fig. E*, from a centre, or they may be described as at *fig. F*, of the same plate, as you see on the other side at *A* and *C* by ordinates; but the first is by far the easiest method for practice, for if you stick a pin or brad awl in *g*, at *D*, and lay a chalk line to it, you may strike all the radical lines *g* 1, *g* 2, *g* 3, *g* 4, &c., in much less time than the parallel lines in *A* and *C* can be drawn, and with much greater accuracy; and the divisions upon *c n* of the arch *F*, may be marked upon a rod, and readily transferred to the arches *D* and *E*, on *m p*, and *f g*: then move your brad awl out of *g*, and stick it in the crown at *f*, and strike lines from the divisions of *m p* to cross the other lines, will give the points through which the arch must pass; but the reader must recollect that four or five points will not be sufficient in the practice for tracing the curve with accuracy, and therefore a greater number must be found. At the other end of the groin is shown the manner in which it may be fixed, sufficiently intelligible for a workman.

Plate 22.

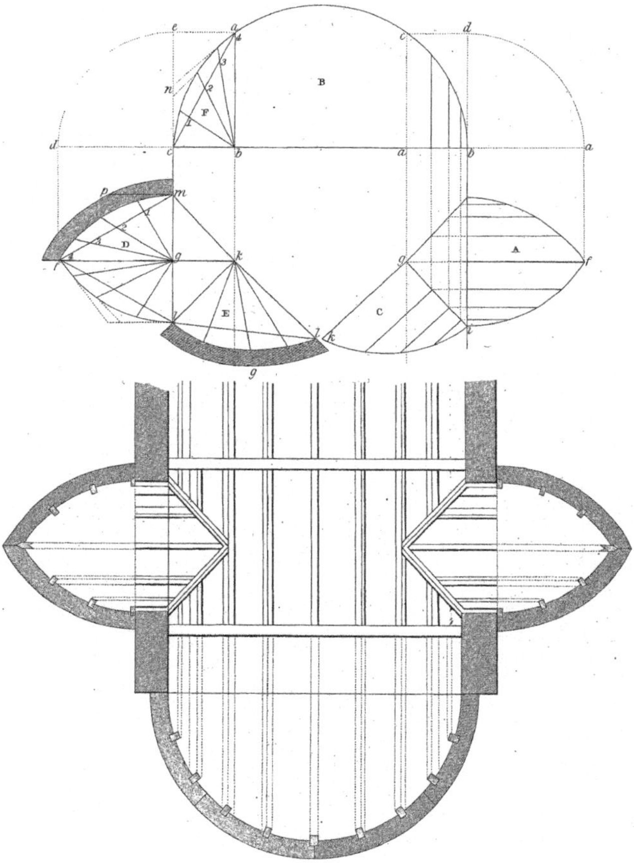

Plate 23.

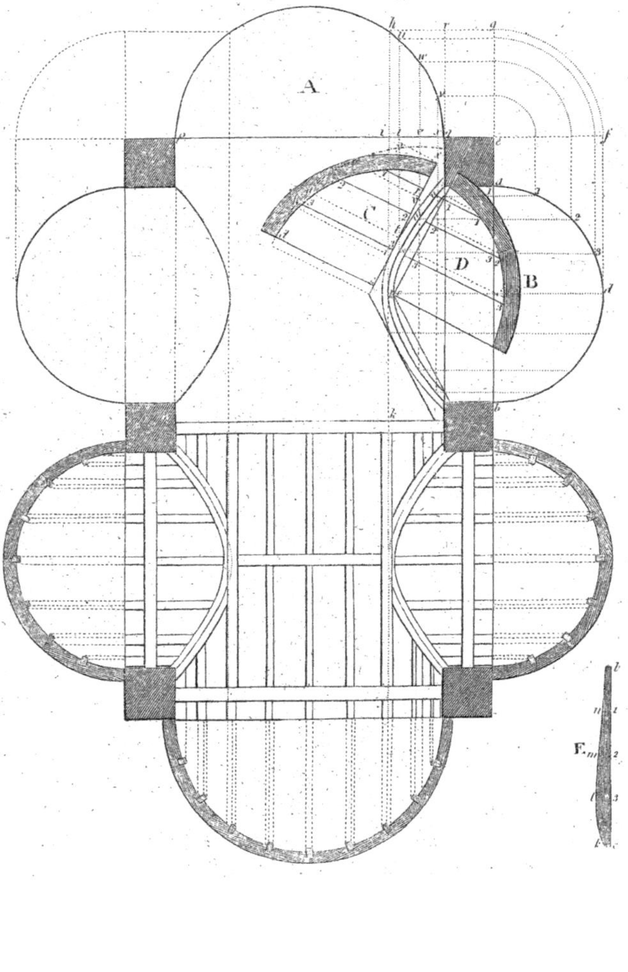

PLATE 23.

A CYLINDRO-CYLINDRIC ARCH.

DEFINITION.

A cylindro-cylindric arch or Welsh groin is an under pitch groin, whose side and body arches are both given semicircles, or they may be similar segments of circles cutting through one another, whose intersections do not meet in a plane surface, that is, the place of the ribs will not be straight upon the plan, but will generate a curve line.

Given the Body Rib A, *and the side Rib* B, *of a cylindro-Cylindric Arch, to find a Mould for the intersecting Ribs.*

Divide half the arch *B*, into any number of equal parts, 1, 2, 3 *d*, or they may be taken at discretion, and from these points let fall perpendiculars to *a b*, its base; produce them at pleasure; also from the same points 1, 2, 3, *d*, draw lines parallel to *a b*, the base of *B*, to intersect the perpendicular line *e f*; transfer the divisions from *e f* to *e g*; then from the division of *e g* draw lines parallel to *p q*, to intersect the body rib *A* at the points *h u w y*: from these points draw perpendiculars to *p q*, its base, and continue them to intersect with the perpendiculars from *B*, at the points *k*, *l*, *m*, *n*, between *C* and *D*; then trace a curve through these points, which will be the place of the intersecting ribs upon the plan; then draw two other curve lines on each side of *k l m n*, &c., to make the thickness of the rib upon the plan: on the inside of the curve draw two chords for each half to their extremities, draw two other lines parallel to them to touch the outside curve, then the distance between those two straight lines will show what thickness of stuff it will take to make the intersecting rib; through the points *k l m n*, &c., draw perpendicular lines to the chords, make the heights *c d*, 3 3, 2 2, 1 1, &c., at *D*, equal to the corresponding heights at *B*: then *D* is the mould for the intersecting rib; *C* is the same as *D*.

To range the Ribs, so that they will stand perpendicular over the Plan.

At the points *x*, *v*, *t*, *i*, in the base of *C*, draw the parallel dotted lines to the ordinates of *C* and *D*, and make their corresponding heights equal to those of the arch *B* or *A*; draw the dotted curve line *h u w y* at *C*, and it will show how much is to be beveled off on that side of the rib; in like manner the other side *D* is beveled, as is shown by the dotted curve line.

To find a Mould to bend under the intersecting Ribs, so that it shall give the Place of the Angle truly upon the Plan.

Take the stretch round the under side of the rib *D* at the dots, by bending a thin slip of wood round it, mark it at each dot, and stretch it out along the straight line *b c* at *E*, draw the ordinates across, and prick them from the plan that lies between *D* and *C*, then *E* agreeably to the letters will be the mould required.

Note. The straight edge of the mould must be kept exactly to the face of the rib; when it is bent round, then draw a curve round the under side of the rib by the other edge of the mould, will give the true place of the angle.

PLATE 24.

There will be no occasion for explaining the lines of this groin, as they are of the same nature as those in the last plate; but it will be proper to take notice, that this is a bevel groin; the ribs must lie in the same direction as the plan of the groin, which will make them longer than their corresponding given arches at the top, but of the same height; they are consequently ellipses, being the sections of cylinders; therefore, to make a rib over *l m*, across the two piers, take the extent of the base *l m*, and the height of the given arch *n o*, and describe an ellipsis; and to describe the side arches between any two piers, as from *a* to *b*, take the extent *a b*, and the height of the given arch, *p g*, at *A*, and describe an ellipsis, it will give the proper form of the rib to stand over *a b*; the intersecting ribs will require two moulds *C* and *D*, owing to the groins being bevel upon the plan.

Note. The letters are marked the same upon *D* and *C* as they are upon *E*, to show they are traced from it.

Plate 24.

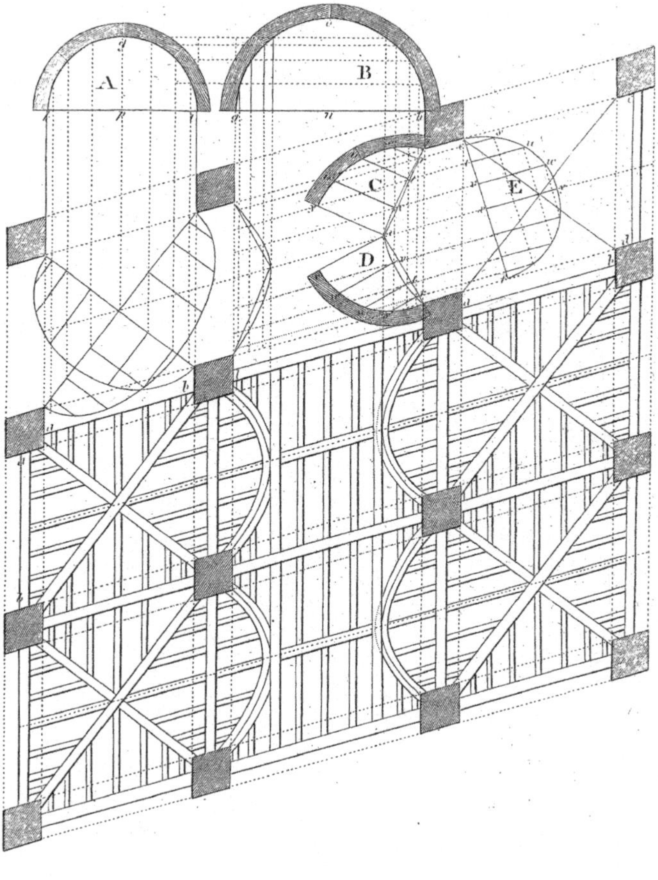

Plate 25.

Fig. 1.

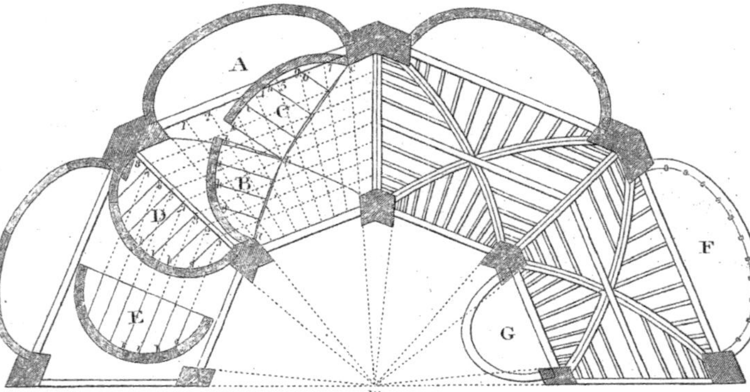

Fig. 2.

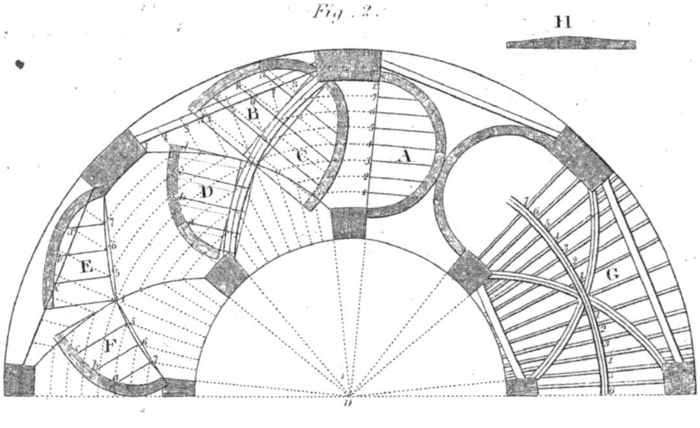

PLATE 25.

To describe the intersecting or Angle Ribs of a Groin standing upon an octagon Plan, the Side and Body Ribs being given both to the same Height.

FIG. 1. *E* is a given body rib, which may be either a semicircle or a semi-ellipsis, and *A* is a side rib given of the same height; *D* is a rib across the angles: trace from *E*, the basis of both being divided into a like number of equal parts, divide the base of the given rib *A*, into the same number of parts; from these points draw lines across the groin to its centre at *m*, and from the divisions of the base of the other rib *D*, draw lines parallel to the side of the groin, then trace the angle curves through the quadrilaterals, will be the place of the intersecting ribs; draw the chords *a b* and *b c*, then prick the moulds *B* and *C* from *E* or *D*, but take care not to prick them from the crooked line at the base, but from the straight chords *a b* and *b c*.

To describe and range the Angle Ribs of a Groin circular upon the Plan, the Side and Body Arches being given, as in the last Groin.

The ribs are described in the same manner as in the last example for the octagon groin, or in the same manner as the cylindro-cylindric, Plate 23, and the ranging is found in the same manner as is described in that plate.

Note. *E* and *F* are the same moulds as are shown at *B* and *D*.

PLATE 26.

The Side Rib A, *and the Angles being given straight upon the Plan, to find the Angle Rib* G, *and the Body Rib* C.

Let the rib *A* be supposed to be placed over the straight line *a b*, as its base, which divide into any number of equal parts, as eight, from the points of division draw lines to the centre of the groin to intersect the angles at *a, b, c, d, e, f g*, these points will give the perpendiculars of the ordinates of *G*, which, being made respectively equal to those of *A*, will give the curve of the rib *G*. If from the points *a, b, c*, &c., arcs be drawn from the centre of the groin to intersect the base of *C*, at 1, 2, 3, 4, 3, 2, 1, and perpendiculars be drawn and made correspondingly equal to those of *A*, and *C* be traced through these points, then *C* will be the body rib.

How to describe the Ribs of a Groin over Stairs upon a circular Plan, the body Rib being given.

FIG. 2. Take the tread of as many steps as you please, suppose nine, from *E*, and the heights corresponding to them, which lay down at *F*; draw the plan of the angles as in the other groins, and take the stretch round the middle of the steps at *E*, and lay it from *a* to *b* at *F*; make *d e* perpendicular to *d c* at *B*, equal to *d e* at *F*, draw the hypothenuse *e c*, draw perpendiculars from *d c* up to *B*, and prick *B* from *A*, as the figures direct, then *B* is the mould to stand over *a b*; draw the chords *a* 4 and 4 *m* at the angles, make *a* 9, 4 *h*, perpendicular to them, each equal to half the height *d e*, at *B* or *F*, draw the hypothenuse *g* 4, and *h m*, draw the perpendicular ordinates from the chords through the intersection of the other lines that meet at the angles, then trace the moulds *D* and *C*, from the given rib *A*, will form the moulds for the angle or intersecting ribs.

Note. The reason that the angle ribs *D* and *C* are laid contrary ways, is only to avoid confusion.

Plate 26.

Fig. 1.

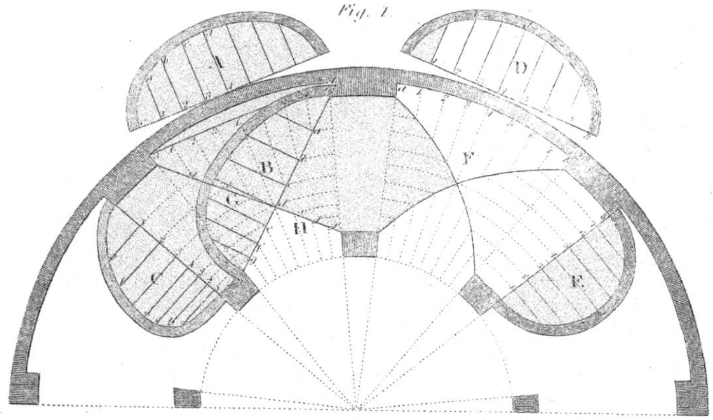

Fig. 2.

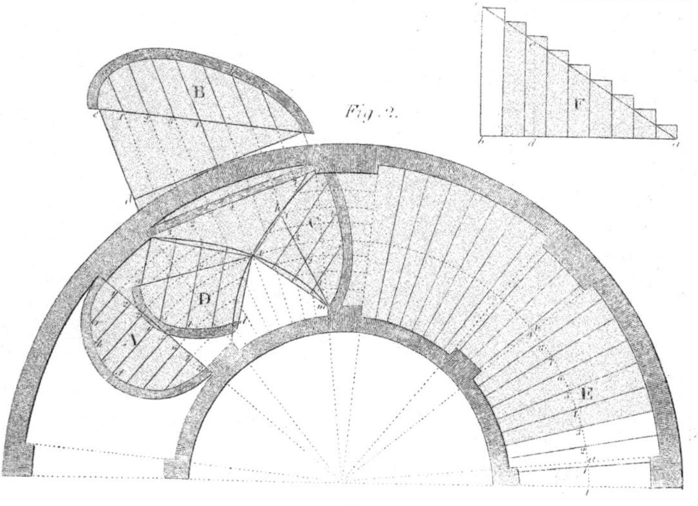

Plate 27.

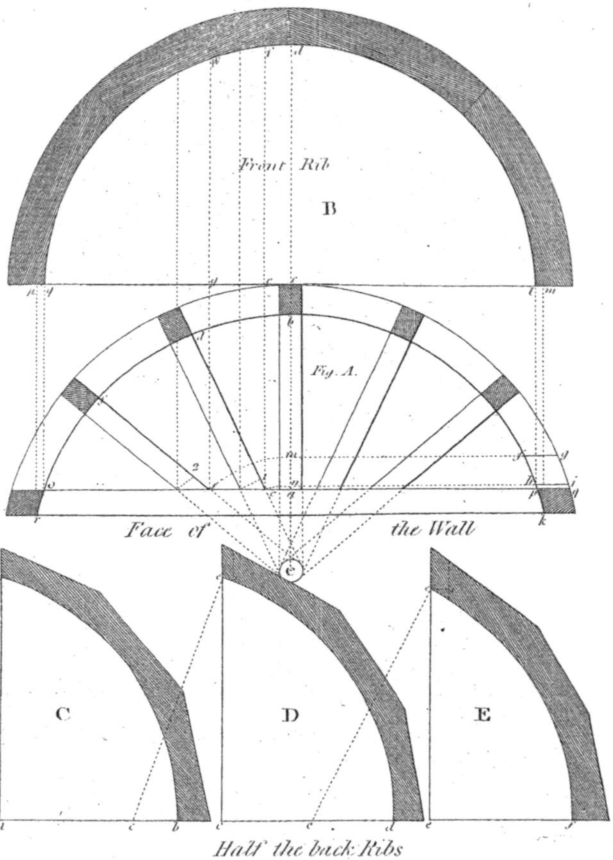

Half the back Ribs

PLATE 27.

As all the sections of a sphere are circles, and those passing through its centre are equal, and the greatest which can be formed by cutting the sphere: it is therefore evident that if the head of a niche is intended to form a spherical surface, the most eligible method is to make the plane of the back ribs pass through the centre; this may be done in an infinite variety of positions; but perhaps the best, and that which would be easiest understood, is to dispose them in vertical planes. If the head is a quarter of a sphere, the front rib, and the still plate or springing, on which the back ribs stand, will curve equally with the vertical ones; but if otherwise, they will be portions of less circles. But it is evident if the front and springing ribs are intended to be arcs less than those of semicircles, either equal to each other or unequal, that as they are posited at right angles to each other, there can be only one sphere which can pass through them; consequently if the places of the vertical ribs are marked on the plan, these ribs can have only one curve: in the former case no diagram is necessary, but in the latter it may be proper to show how the vertical ribs and their situation on the front rib are found.

To get out the Ribs for the Head.

From the centre *C* draw the ground-plan of the ribs as at *figure A*, and set out as many ribs upon the plan as you intend to have in the head of the niche, and draw them all out towards the centre at *C*. Place the foot of your compass in the centre *C*, and from the ends of each rib, at *e* and *c*, draw the small concentric dotted circles round to the centre rib at *m* and *n*; and draw *m g* and *n i*, parallel to *r k*, the face of the wall; then from *q* round to *c* upon the plan, is the length and sweep of the centre rib, to stand over *a b*; and from *i* round to *e*, the length and sweep of the rib that stands from *c* to *d* upon the plan; and from *g* round to *e* is the sweep of the shortest rib, that stands from *e* to *f* upon the plan.

Secondly. To bevel the Ends of the Back Ribs against the Front Rib.

The back ribs are laid down distinct by themselves at *C*, *D* and *E* from the plan. Take *c* 1, in *figure A*, and set it from *c* to 1 in *D*, will give the bevel of the top of the rib *D*. And from *figure A*, take from *e* to 2 upon the plan, and set from *e* to 2 in the rib *E*, will give the bevel of the top.

Thirdly. To find the places of the Back Ribs where they are fixed upon the Front.

From the points *a*, *c*, and *e*, at the ends of the ribs, in the plan, *figure A*, draw the dotted lines up to the front rib, to *d f* and *w*, which will show where they are to be fixed upon the front rib. The double circle upon the front rib shows the ranging.

PLATE 28.

To find the Curve of the Ribs of a Spherical Niche, the Plan and Elevation being given Segments of Circles.

In *fig. A* is the elevation of the niche, being the segment of a circle whose centre is *t*; at *B* is the plan of the same width, and may be made to any depth, according to the place it is intended for, and its centre is *c*; on the plan *B*, lay out as many ribs as it will require, draw them all tending to the centre at *c*, they will cut the plan of the front rib in *g, f, e, d*; through the centre *c*, draw the line *m n*, parallel to *a b*, the plan of the front rib; put the foot of your compass in the centre at *c*, draw the circular lines from *a, g, f, e, d*, to the line *m n*, and make *c s* equal to *u t*, that is, make the distance from the middle of the chord line *m n* to *s*, the centre of the arch at *C*, equal to the distance from the middle of the chord at the top at *fig. A*, to its centre at *t*; then place the foot of your compass in *s*, as a centre, and from the extremities *m* or *n*, describe the arch at *C*; with the same centre draw another line parallel to it, to any breadth as you intend your ribs shall be; then *C* is the true sweep of all the back ribs in the niche.

Note. The points *l, k, i, h*, show what length of each rib will be sufficient from the point *m*; from *h* to *m* is the rib that will stand over *d x*, from *i* to *m* is the rib that will stand over *e w*, from *k* to *m* over *f v*, and from *l* to *m* over *g w*: the other half is the same.

Through the centre *t*, draw D E, parallel to *a b*, complete the semicircle E F G D, then D E is the diameter; through *n* draw *n* A parallel to *u d*, in the centre *t*, with the distance *t* A describe another semicircle whose diameter is C B; then will the semicircle C E G A B be equal to a vertical section of the globe, standing on K I, passing through its centre at *c*, which is the same curve as the rib at *C*, because *u* A is equal to *c n*, and *c s* bisecting *m n* at right angles, is equal to *t u*, bisecting E A at right angles: therefore the hypothenuse *t* A, that is, the radius of the circle B A G E C, is equal to *s m* or *s n*, the radius of the circle or rib at *C*.

Plate 28.

Fig. 1.

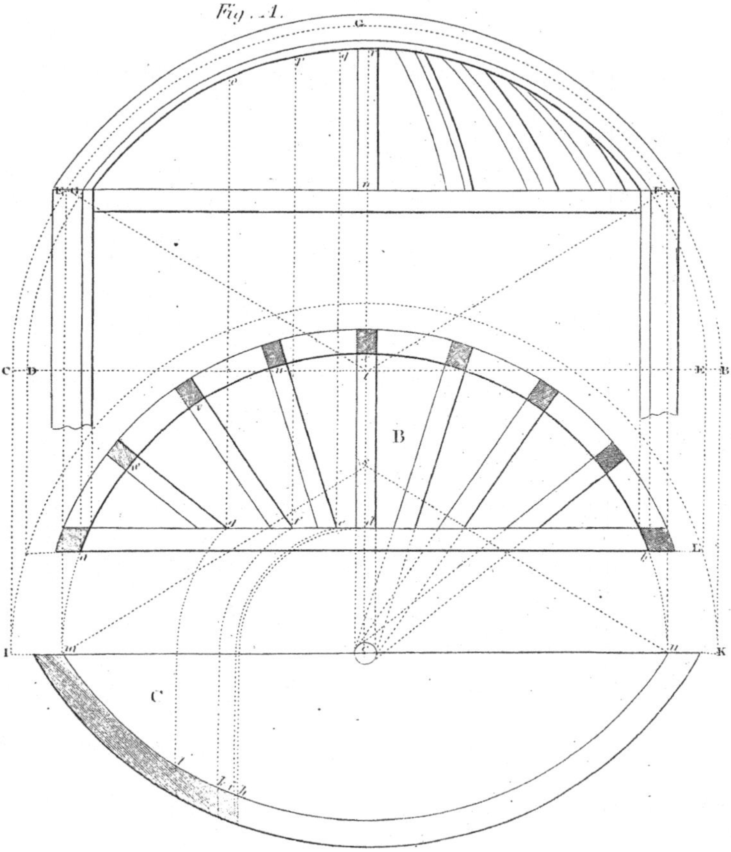

Plate 29.

Fig. 1.

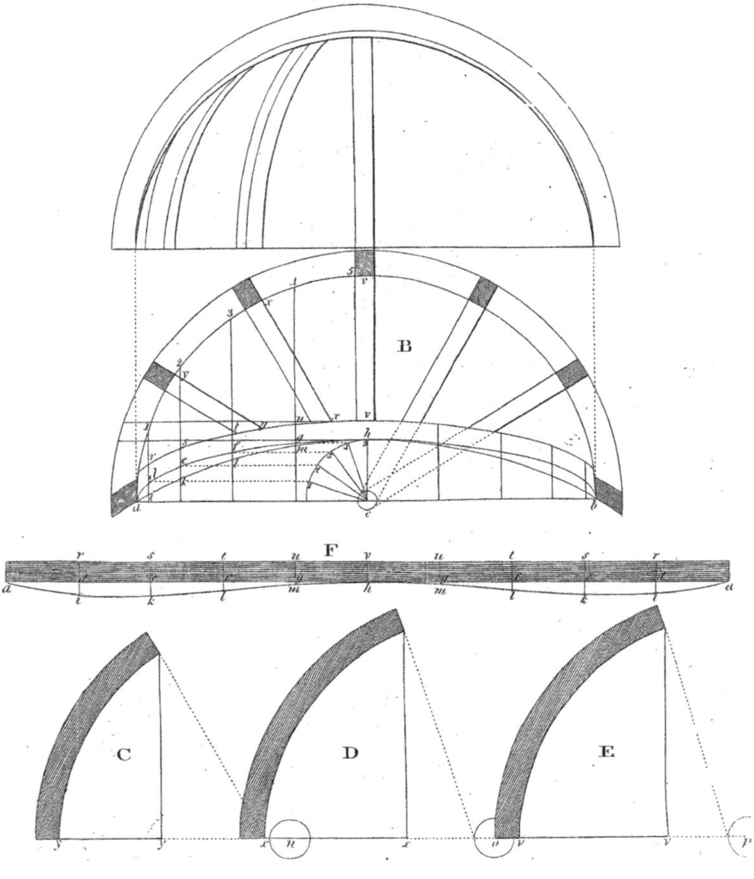

PLATE 29.

The Plan of a Niche in a circular Wall being given, to find the Front Rib.

B is the plan given, which is a semicircle whose diameter is *a b*, and *a, i, k, l, m, h*, the front of the circular wall; suppose the semicircle *B*, to be turned round its diameter *a b*, so that the point *v* may stand perpendicular over *h* in the front of the wall, the seat of the semicircle standing in this position upon the plan will be an ellipsis; therefore divide half the arch of *B* upon the plan into any number of equal parts, as five; draw the perpendiculars 1 *d*, 2 *e*, 3 *f*, 4 *g*, 5 *h*, upon the centre *c* with the radius *c h*, describe the quadrant of a smaller circle, which divide into the same number of equal parts as are round *B*; through the points 1, 2, 3, 4, 5, draw parallel lines to *a b*, to intersect the others or the points *d, e, f, g, h*, through these points draw a curve, it will be an ellipsis; then take the stretch-out of the rib *B*, round 1, 2, 3, 4, 5, and lay the divisions from the centre both ways at *F*, stretched out; take the same distances *d i, e k, f l, g m*, from the plan, and at *F* make *d i, e k, f l*, equal to them, which will give a mould to bend under the front rib, so that the edge of the front rib will be perpendicular to *a, i, k, l, m*.

Note. The curve of the front rib is a semicircle, the same as the ground-plan, and the back ribs at *C D* and *E* are likewise of the same sweep.

The reason of this is easily conceived, the niche being part of a globe, the curvature must be everywhere the same, and consequently the ribs must fit upon that curvature.

Note. The curve of the mould *F* will not be exactly true, as the distances *d i, e k, f l*, &c., are rather too short for the same corresponding distances upon the soffit at *F*, but in practice it will be sufficiently near for plaster work; but those who would wish to see a method more exact, may examine Plate 15, *fig. A*, where *C* is the exact soffit that will bend over its plan at *B*.

In applying the mould *F* when bent round the under edge of the front rib, the straight side of the mould *F* must be kept close to the back edge of the front rib, and the rib being drawn by the other edge of the mould, will give its place over the plan.

6

PLATE 30.

The Plan and Elevation of an Elliptic Niche being given, to find the Curve of the Ribs.

Fig. *A*. Describe every rib with a trammel, by taking the extent of each base from the plan whereon the ribs stand to its centre, and the height of each rib to the height of the top of the niche, it will give the true sweep of each rib.

To back the Ribs of the Niche.

There will be no occasion for making any moulds for these ribs, but make the ribs themselves; then there will be two ribs of each kind; take the small distances 1 *e*, 2 *d*, from the plan at *B*, and put it to the bottom of the ribs *D* and *E*, from *d* to 2, and *e* to 1; then the ranging may be drawn off by the other corresponding rib; or with the trammel, as for example, at the rib *E*, by moving the centre of the trammel towards *e*, upon the line *e c*, from the centre *c*, equal to the distance 1 *e*, the trammel rod remaining the same as when the inside of the curve was struck.

Given one of the common Ribs of the bracketing of a Cove, to find the Angle Bracket for a rectangular Room. Fig. F.

Let *H* be the common bracket, *b c* its base; draw *b a* perpendicular to *b c*, and equal to it draw the hypothenuse *a c*, which will be the place of the mitre; take any number of ordinates in *H*, perpendicular to *b c*, its base, and continue them to meet the mitre line *a c*, that is, the base of the bracket at *I*; draw the ordinates of *I* at right angles to its base; then the bracket at *I*, being pricked from *H*, as may be seen by the figures, will be the form of the angle rib required.

Note. The angle rib must be ranged either externally or internally, according to the angle of the room.

Having given a common Bracket K, Fig. G, *for Plaster Cornice, to find the Mitre Bracket* L.

Proceed as in the last example, and you will have the bracket required.

Plate 30

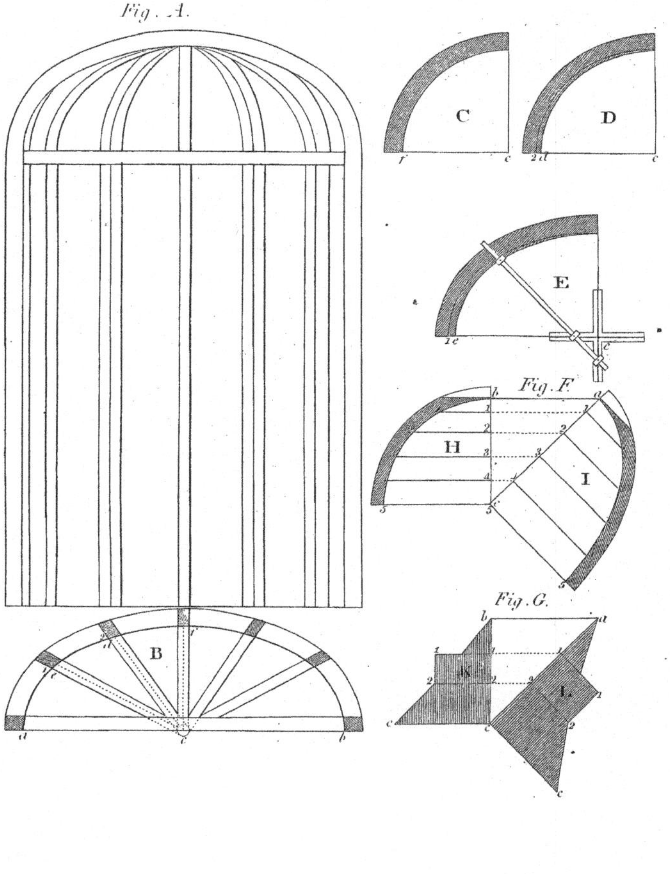

Plate 31.

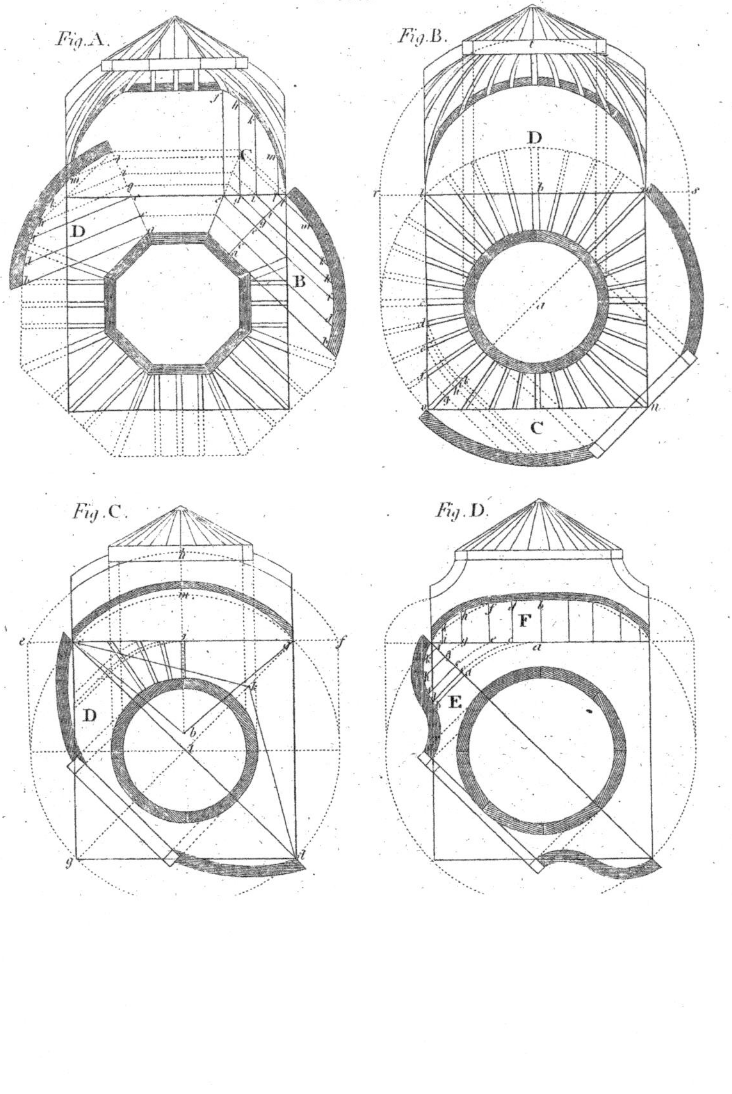

PLATE 31.

OF PENDENTIVES AND INTERIOR DOMES WHEN PLACED OVER THE OPENINGS OF ROOMS.

One of the Ribs of a Dome being given, and the Plan of the Opening of a Staircase which is square, and an octagon Curb at the Top for a Sky-light; to find the Ribs and the springing Curve on each Side of the opening of the Staircase, where the Foot of the Ribs come, so that Part of the Dome shall be an octagon finish, agreeably to the Curb.

FIG. *A.* Let *B* be the given rib; take any number of perpendicular ordinates to its base at pleasure, from the points *a, c, e, g, i, l,* where they intersect its base, draw parallel lines to the sides of the curb, returning round each diagonal, if there is more than one, till it cut the base of the angle rib *D*; at the points *a, c, e, g, i, l,* draw the ordinates of *D,* and prick it from *B,* will be the angle rib; and at the points *e, g, i, l,* at *C,* upon the side of the opening of the staircase, draw the perpendicular ordinates, and prick *C* from *B,* agreeably to the letters; then the curve *C* will be the true place for the foot of the ribs upon the side of the staircase, and the part that lies in the middle is a straight line parallel to the horizon.

The Opening or Plan of the Room being a Square as before, and the vertical Section of a semicircular, to find the springing Curve D *on the Side of the Room for the Foot of the Ribs, so that it shall finish to a circular Curb at the Top.*

On the side of the staircase *l m,* as a diameter, describe a semicircle; *D* will be the true place for the foot of the ribs; this is evident, for every section of a semi-globe, at right angles to its base, is a semicircle, and this is the same thing if truly considered.

Note. All the ribs of this dome are cut by the rib at *C,* as explained by the perpendicular lines; draw round the centre *a,* from the points of each bracket, at *c d e f,* to the points *k i h g,* from these points draw perpendicular dotted lines, and these will show what length each bracket must have according to its place.

The vertical Section of a Segment Dome passing through its Centre being given, the Plan of the Opening of the Room being still a Square, as before, to find the Section upon each Wall for the Springing of the Ribs, to finish to a circular Curb at the Top.

Let the section *D* across the angle be given, whose centre is *k,* and the distance of the centre from the chord *k l*; bisect the side *c g* of the wall *b h,* at right angles at the point *i*; from *i* make *i b* equal to *l k*; with a radius *b g* or *b c,* describe the segment *c, m g* will be the true place of springing of the ribs; all the other lesser ribs are cut from the angle rib *D*: all this is evident from the sections of a globe, which is already described in the Geometry.

FIG. *D* is of the same nature as the others, having an ogee top; the section *F* is traced from *E.*

PLATE 32.

FIG. *A* is the plan of an elliptical domical sky-light over a staircase; *B* and *C* are the sections, which show how to place your ribs.

How to proportionate the Length of the inside Curb to any Width given.

Proceed as directed in page 48 for an elliptical dome, that will determine the true length to the width.

How to proportionate the circumscribing Ellipsis, to pass through the Angles at a, b, c, *and* d, *to have the same Proportion as* a b, *and* b c, *of the Sides of the Staircase.*

Proceed as directed in *fig.* 5, Plate 6, in the Geometry.

To describe the Ribs.

The rib over from *n*, to the centre of the trammel in *fig. A*, is a given quarter of a circle, as is shown at *F*, and of course all the other ribs must come to the same height with it. Suppose it was required to find a rib over *d p*, you must take the full extent from *d* to the centre, and describe the quarter of an ellipsis *D*; then the part over *d p* will be as much of it as is wanted: in the same manner *E* will be described, and the part over *i o* is what is wanted of this rib; the same letters are marked upon the bases of *D* and *E*, as they are in the plan *fig. A*. Every other rib is described in the same manner.

To find the Springings on each Side of the Room for the Foot of the Ribs to stand upon.

Describe the semicircle *C*, to *c b*, the width of the room, and it will give the bottom of the ribs on that side; and describe a quarter of the ellipsis *B*, for the bottom of the ribs on the other side, to the same height as *C*.

This method depends on this principle, that all the parallel sections of a spheroid are similar figures: therefore a vertical section standing upon *a b*, will be similar to a vertical section passing through its centre; both will be similar ellipses; but *a b* is an ordinate to the conjugate axis, and *b c* is an ordinate to the transverse of the circumscribing ellipsis; by construction half the length of the parallelogram is to half the length of the ellipsis, as half the width of the parallelogram is to half the width of the ellipsis, and a spheroid may be supposed to be generated by the revolution of a semi-ellipsis about its axis; hence it follows, that all sections of a spheroid parallel to the axis are similar figures, consequently the section *B* is similar to the circumscribing ellipsis of the ground plan.

Plate 32.

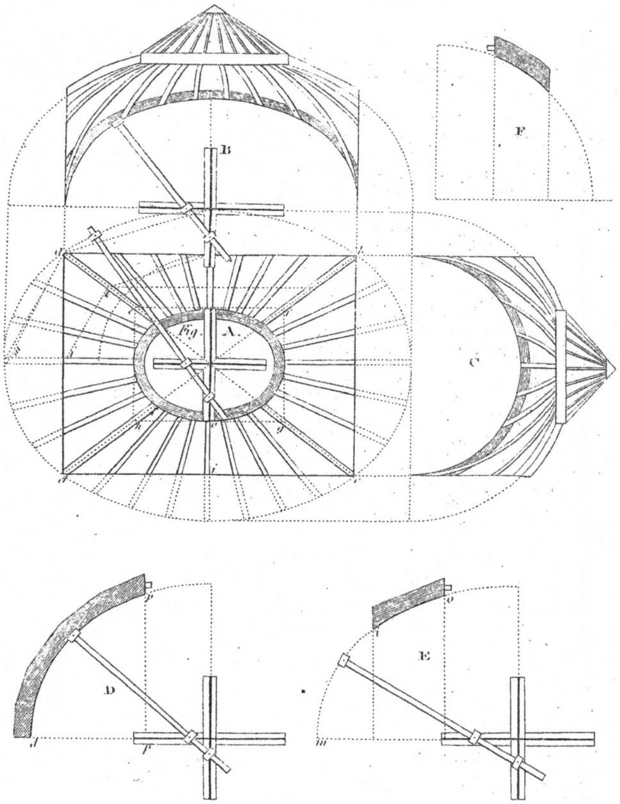

Plate 33.

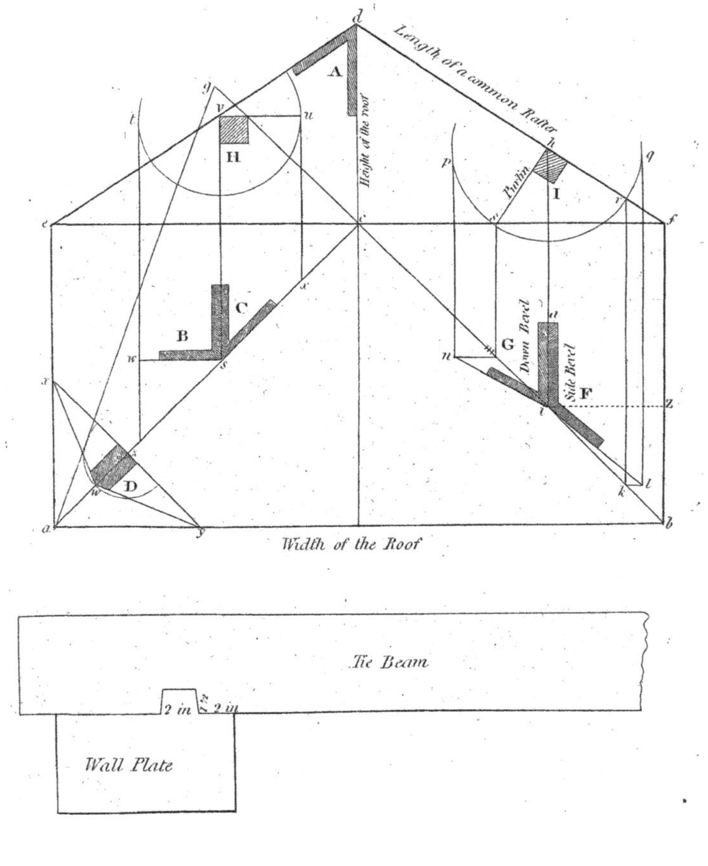

Length of a common Rafter
Height of the roof
Purlin
Down Bevel
Side Bevel
Width of the Roof
Tie Beam
2 in
2 in
Wall Plate
A
B
C
D
F
G
H
I

PLATE 33.

Let *a b* be the end of a rectangular roof, *a e* and *b f* being a part of each side, let *a e* and *b f* be each equal to half the width *a b* of the roof; join *e f*; bisect *e f* in *c*; draw *c d* perpendicular to *e f*, and make *c d* equal to the height of the roof; join *d e* and *d f*, and *d e* and *d f* are the length of the principal rafters; join *a c* and *b c*; produce either diagonal, as *b c* to *g*, make *c g* equal to *c d*; join *a g*, and *a g* is the length of each hip.

Draw any line *x y* perpendicular to the seat *a c* of a hip, cutting *a e* and *a b* at *x* and *y*, and *a c* at *z*: from *z* describe a tanged circle to *a g*, cutting *a c* at *w*; join *w x* and *w y*, and the angle *y w x* is the inclination of the planes which form the hip angle, and is what is generally termed the backing of the hip.

In this plate one end of the roof is shown in order to show two cases: the first is when the purline lies level, or having two sides parallel to the horizon; the square at *B*, and the bevel at *C*, will show how to draw the end of the purline in this easy case; but the following method is universal in all positions of the purline.

Note. There will be no occasion to draw this at large; as the bevels will be the same if done to ever so small a scale, and the sides may be measured from the scale.

To find the Bevels of a Purline against the Hip Rafter.

Let the purline be in any place of the rafter, as at *I*, and in its most common position, that is, to stand square, or at right angles to the rafter; and from the point *h* as a centre, with any radius describe a circle. Draw two lines *q l* and *p n*, to touch the circle in *p* and *q*, parallel to *f b*; and at the points *s* and *r*, where the circle and two sides of the purline intersect, draw two parallel lines to the former, to cut the diagonal in *m* and *k*; and draw *m n* and *k l* perpendicular to *s m* and *r k*, and join the points *n*, *i* and *k*, *i*; then *G* is the down bevel, and *F* the side bevel of a purline: these two bevels, when applied to the end of the purline, and when cut by them, will exactly fit the side hip rafter.

To find the Bevels of a Jack Rafter against the Hip.

By turning the stock of the side bevel of the purline, at *F*, from *a* round to the line *i z*, will give the side bevel of the jack rafter. And the bevel at *A*, that is, the top of a common rafter, is the down bevel of the jack rafter.

At the bottom is shown the manner of cocking down the tie beam upon the wall plate; the proper size of the cocking is figured.

PLATE 34.

This plate shows the manner of framing a roof in ledgement; but as roofs are seldom executed in this manner, I shall not be very particular in describing its lines. The following description for winding will serve for any.

Plate 34.

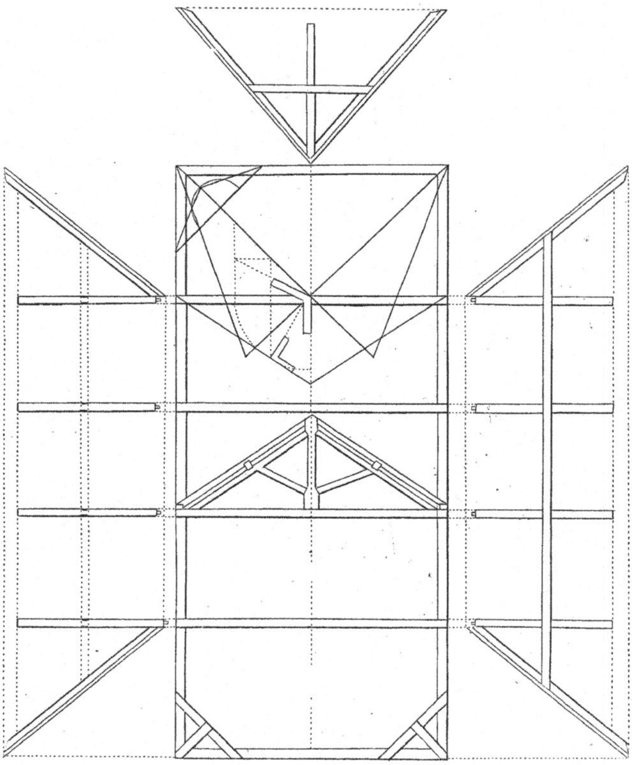

Plate 35.

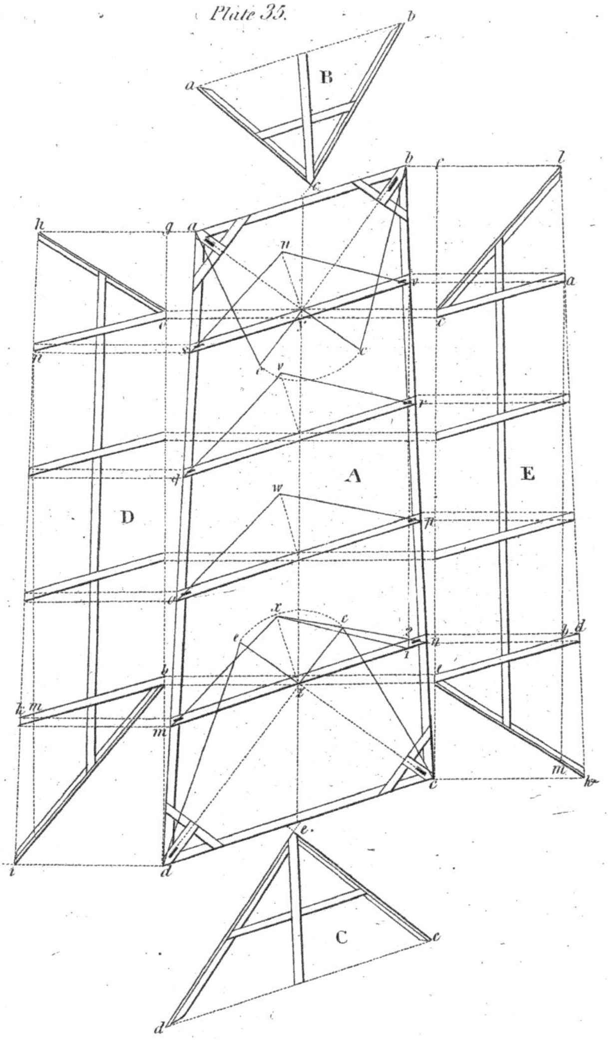

PLATE 35.

How to lay out an irregular Roof in Ledgement, with all its Beams lying bevel upon the Plan, so that the Ridge may be level when finished; the Plan and Height of the Room being given.

The lengths of the common and hip rafters are found as usual. From each side in the broadest end of the roof, through *c* and *d*, draw two parallel lines to the ridge line; draw lines from the centres and ends of the beams perpendicular to the ridge line, and lay out the two sides of the roof *D* and *E*, by making *e d* at *E*, equal to *x n* in *A*, the length of the longest common rafter and *c a* in *E*, equal to *u v* at *A*, and so on with all the other rafters.

To find the Winding of this Roof.

Take *y v* half the base of the shortest rafter: and apply this to the base of the longest rafter from *z* to 1; then the distance from 1 to 2 shows the quantity of winding.

How to lay the Sides in Winding.

Lay a straight beam along the top ends of the rafters at *E*, that is, from *c* to *e*, and lay another beam along the line *a b*, parallel to it, to take the ends of the hip rafters of *m* and *l*, and the beams to be made out of winding at first. Raise the beam that lies from *a* to *b*, at the point *b*, to the distance 1 2 above the level; which beam, being thus raised, will raise all the ends of the rafters gradually, the same as they would be when in their places.

The same is to be understood of the other side *D*; the ends are laid down in the same manner as making a triangle of any three dimensions.

To satisfy the curious, I have given the lines of this roof; but in practice there is not the least occasion for framing the sides in winding; for, instead of the ridge line, the top is made level at the widest end of the roof, from the narrowest end, which begins at a point; and by this means the sides may be framed quite out of winding, which will have a much better effect than any winding roof can have.

PLATE 36.

POLYGON ROOFS.

The methods of constructing regular polygons upon any given side, are shown at *figs*. 4, 5, and 6, in Geometry.

The Plan of a Polygon Roof being given, and one of the common ribs standing upon that Plan, to find the Angle Rib, and the Form of the Boards that will cover it when the Ribs are fitted up.

In *fig*. *A* let *B* be the given rib; divide the curve in any number of equal parts, as four, and lay them at *D* from *a* to 4, which bisects *b b*, the side of the polygon, at right angles; through these points draw lines parallel to the side *b b* of the polygon; at *B* and *D* make 1 *c* at *D* equal to *c c*, between *B* and *C* make 2 *d* equal to *d d*, and 3 *e* equal to *e e*, &c., and through the points *b*, *c*, *d*, *e*, *f*, draw a curve line, which will be the form of the boarding; from the points *g*, *f*, *e*, *d*, *c*, draw lines at right angles to *g b*, the base of the angle rib, and prick the rib *C* from *B*, as they are marked by the letters, which is plain.

Note. The more parts there are in this operation, the truer will it be, or any other of this nature.

In the same manner may the covering and angle ribs of any other polygon be found, whatever may be the form of the ribs, as is shown at figures *B* and *C*.

To find the Covering of a spherical Dome.

Fig. *D*. Make a circle *i c k f*, of the plan of the dome, and if it is a semi-globe, take the stretch-out of one quarter for the length of a board; make the length of *K* from *a* to 4 equal to it, and let *c c*, at the bottom, be any breadth that the board will admit of; on the base *c c* as a diameter, make a semircle; divide half the arch line into any number of equal parts; draw the little lines 1 1, 2 2, 3 3, parallel to *c c*, the base of the board, and divide the height into the same number of equal parts; draw the ordinates across; make 1 1, 2 2, 3 3, upon these ordinates, equal to 1 1, 2 2, 3 3, in the semicircle at the bottom; a curve being drawn through these points will be the mould *K* for the covering.

To cover a spherical Dome when the Top does not rise so high as a Semicircle, but only a Segment.

Suppose *l d* to be the height of the dome at *F*, and the width *c f* of the dome as before, upon the chord *c f*, with the perpendicular height *l d* describe a segment, which will be the same as a vertical section standing upon *c f*; here is only one half of the segment, which is sufficient: draw the chord *c d*; take *c a* equal to half the width of a board, whatever it will admit of: draw *a b* perpendicular to cut the chord *c d* at *b*; take the stretch or circumference of the arch *c d*, and make the length of *I* from *a* to 4 equal to it; take the double of *a c*, at *F*, and make it the base of the board at *I*; take *a b* from *F*, and set it upon the base of *I*, upon the middle of *c c* from *a* to *b*; and with the chord *c c*, and the height *a b*, describe a segment upon the bottom of the board at *I*; divide one-half into any number of equal parts; likewise divide the height of the board *I* into the same number of equal parts; draw ordinates in both, and the board *I* will be completed, as in the same manner as that of *H*, described before.

Plate 36.

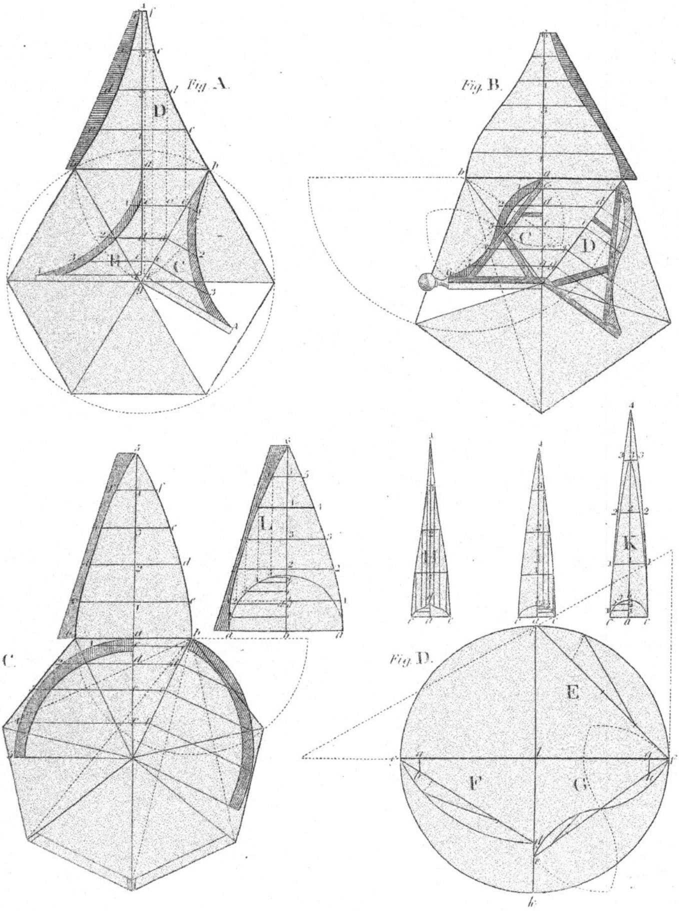

Plate 37.

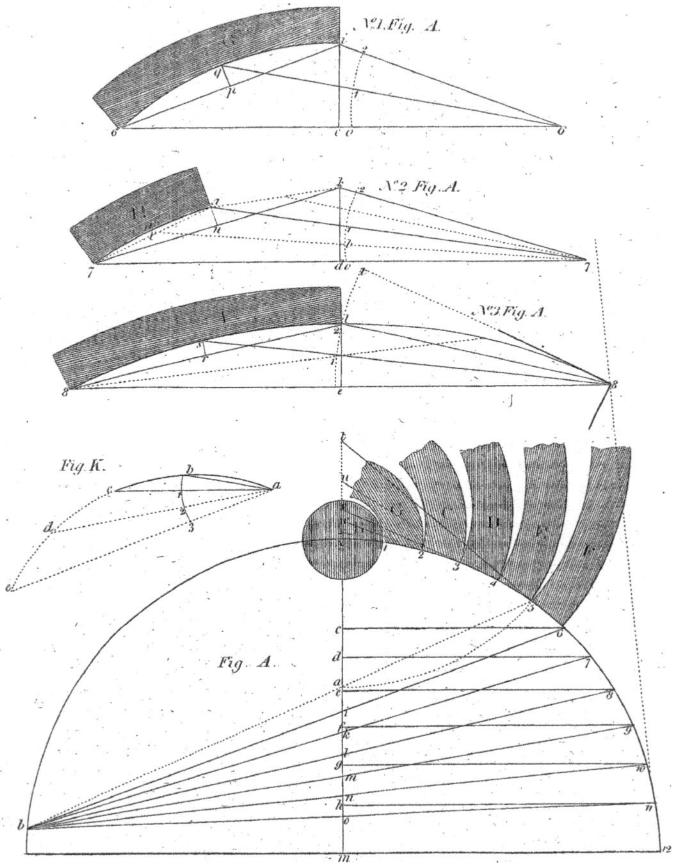

PLATE 37.

CIRCULAR DOMES.

As the common method of finding the centres for describing the boards to cover a horizontal dome, will be found in practice very inconvenient, for those boards which come near to the bottom, I shall in this place show how to remedy that inconvenience.

To find the Sweep of the Boards on the Top. FIG. A.

Divide round the circumference of the dome into equal parts at 1, 2, 3, 4, 5, 6, &c., each division to the width of a board, making proper allowance for the camber of each board; draw a line through the points 1, 2, to meet the axis of the dome at *x*; on *x*, as a centre, with the radii *x* 1 and *x* 2 describe the two concentric circles, it will form the board *G*; in the same manner continue a line through the points 2 and 3 at *C*, to meet the axis in *w*; then *w* is the centre for the board *C*; proceed in the same manner for the boards *D*, *E*, and *F*.

Now suppose *F* to be the last board that you can conveniently find a centre, for want of room; on *t* its centre, and the radius *t* 5, make from *t* on the axis of the dome *t a*, equal to *t* 5; through the points 5 and *a* draw the dotted line 5 *a b*, to cut the other side of the circumference of the dome at *b*; from the points 6, 7, 8, 9, 10, 11, draw radial lines to *b*, to cut the axis of the dome at *i*, *k*, *l*, *m*, *n*, *o*; also through the points 6, 7, 8, 9, 10, 11, draw the parallels 6 *c*, 7 *d*, 8 *e*, &c., then will each of these parallel lines be half the length of a chord line for each board; then take *c* 6 from *fig. A*, which transfer to No. 1, from *c* to 6 and 6; make the height *c i*, at No. 1, equal to *c i*, at *fig. A*; and draw the chords *i* 6 and *i* 6; then upon either point 6, as a centre with any radius, describe an arch of a circle 0 1 2; divide it into two equal parts at 1, and through the points 6 and 1 draw 6 *q*; bisect *i* 6, in *p*; draw *p q* perpendicular; then *i* 6 is the length, and *p q* the height of the board *G*, which may be described as in *fig.* 4, plate 5, of the Geometry. The reader must observe, that the length of a board is of no consequence so as the true sweep is got, which is all that is required. Proceed in the same manner with No. 2, by taking *d* 7 from *fig. A*, and place it at No. 2, on each side of *d* at 7 and 7, and take *d k*, from *fig. A*, and make *d k* at No. 2, equal to it; draw the chords *k* 7 and *k* 7, and bisect *k* 7 at *n*; draw *n a* perpendicular; upon the other extremity at 7, as a centre, describe an arch 0 1 2, and bisect it at 1, and through the points 7 and 1 draw the line 7 *a*, to cut the perpendicular *n a* at *a*; but if the distance *k* 7 is too long for the length of a board, bisect the arch 0 1 at *b*; through 7 and *b* draw 7 *t*, and draw the little chord *a* 7, and bisect it at *t*; draw *t u* perpendicular to intersect 7 4 at *u*; and with the chord 7 *a* and the height *t u*, describe the segment *H*.

In the same manner may the next board *I* be found, and by this means you may bring the sweep of your board into the smallest compass, without having any recourse to the centre.

Suppose it were required to draw a Tangent from 8 *at No.* 3, *without having recourse to the Centre.*

Bisect the arch 8 *l* 8 at *l*; on 8 as a centre, with a radius 8 *l*, describe an arch *e l t*; make *l t* equal to *l e*; draw the tangent *t* 8.

Given three Points in the Circumference of a Circle, to find any Number of equidistant Points beyond those that will be in the same Circumference.

FIG. *K.* Suppose the three points *a*, *b*, *c*, to be given; to one of the extreme points *a* join the other two points *b* and *c* by the lines *a b* and *a c*; with a radius *a b*, and the centre *a*, describe the arch of a circle *b* 1 2 3; then take *b* 1, and set it from 1 to 2, and from 2 to 3; through the points 2 and 3, draw *a d* and *a e*; then take *b c*, put the foot of your compass in *c*, and with the other foot cross the line *a d* at *d*; with the same extent put the foot of your compass in *d*, and with the other foot cross the line *a e* at *e*; in the same manner you may proceed for any number of points whatever.

PLATE 38.

SPHEROIDAL DOMES.

FIG. *A* is the plan of a spheroidal dome; *B* is the longest section, *C* the shortest section; at *a a* in *B*, and *b b* in *C*, shows how to square the purlines, so that one side may be fair with the surface of the dome; the dotted lines from *a a* in *B*, and *b b* in *C*, show how to get the length and width of the purline in *fig. A*; but if the sides of the purline were made to stand perpendicular over the plan, the curve of it would be found in the same manner as before; then it would require no more than half the stuff that the other would, and take only half the time in doing, which is a considerable advantage.

How to proportionate the inside Curve for the Sky-light, so as it shall answer to the Surface of the Dome.

Draw the diagonal *i l* and *k m* in *fig. A*, and let *h e* or *g f* be the width, then *h g* or *e f* will be the true length of the curb; because every section parallel to the base will be proportional to the base.

To find the Ribs for this Dome.

The ribs in this are got in the same manner as the ribs for a niche, as directed in page 41; and if the reader understand that, he must know this.

To find the Form of a Board to stand in any Place of the Dome, in order to be bent up to the Crown.

Divide one quarter of the base of the dome at *D* into three equal parts *a b*, *b c*, *c d*, and suppose you would find a board over *a b c* in the plan; draw *a c*, *b c*, *c c*, and *d c*, to the centre at *c*; then take the triangle *a b c* in *D*, and lay it down at *a b c* in *G*; then draw the line *c* 1 1 1, &c., at right angles to *a b*, and describe a rib *G* to the height of the dome, and the length to the perpendicular of the triangle *a b c*, and divide it into five equal parts, lay them along the line 1 1 1, &c., in *H*, and prick the mould *H* from the triangle *a b c*, as the letters are marked. The board *K* will be found in the same manner.

Note. In the practice, you are to divide one-quarter of this dome into as many parts as you think the breadth of a board will contain; and the boards, when got out by this method, will fit to a very great exactness; this is only into three, that the parts may be clearly seen to learners.

If the boards are got out for one quarter to the lines here laid down, the boards that are in the other three quarters will not require any other lines, for every board in the first quarter will be a mould for three more boards.

Plate 38.

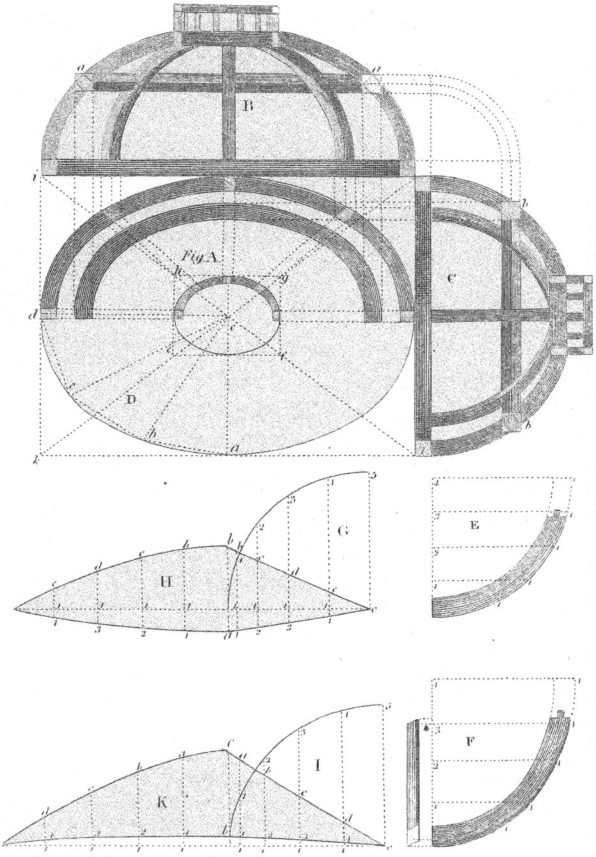

INTRODUCTION

TO

PRACTICAL CARPENTRY.

OF THE COMPARATIVE STRENGTH OF TIMBER.

PROPOSITION I.

THE strengths of the different pieces of timber, each of the same length and thickness, are in proportion to the square of the depth; but if the thickness and depth are both to be considered, then the strength will be in proportion to the square of the depth, multiplied into the thickness; and if all the three dimensions are taken jointly, then the weights that will break each will be in proportion to the square of the depth multiplied into the thickness, and divided by the length; this is proved by the doctrine of mechanics. Hence a true rule will appear for proportioning the strength of timbers to one another.

RULE.

Multiply the square of the depth of each piece of timber into the thickness; and each product being divided by the respective lengths, will give the proportional strength of each.

EXAMPLE.

Suppose three pieces of timber, of the following dimensions:

The first, 6 inches deep, 3 inches thick, and 12 feet long.

The second, 5 inches deep, 4 inches thick, and 8 feet long.

The third, 9 inches deep, 8 inches thick, and 15 feet long. The comparative weight that will break each piece is required.

OPERATIONS.

First.	Second.	Third.
6 deep	5 deep	9 deep
6	5	9
36	25	81
3 thick	4 thick	8 thick
Length 12)108	Length 8)100	Length 15)648(43 and a fifth.
		60
9	12 and a half	
		48
		45
		3

Therefore the weights that will break each are nearly in proportion to the numbers 9, 12, and 43, leaving out the fractions, in which you will observe, that the number 43 is almost 5 times the number 9; therefore the third piece of timber will bear almost 5 times as much weight as the first; and the second piece nearly once and a third the weight of the first piece; because the number 12 is once and a third greater than the number 9.

The timber is supposed to be everywhere of the same texture; otherwise these calculations cannot hold true.

PROPOSITION II.

Given the length, breadth, and depth of a piece of timber; to find the depth of another piece whose length and breadth are given, so that it shall bear the same weight as the first piece, or any number of times more.

RULE.

Multiply the square of the depth of the first piece into its breadth, and divide that product by its length: multiply the quotient by the number of times as you would have the other piece to carry more weight than the first, and multiply that by the length of the last piece, and divide it by its width; out of this last quotient extract the square root, which is the depth required.

EXAMPLE I.

Suppose a piece of timber 12 feet long, 6 inches deep, 4 inches thick; another piece 20 feet long, 5 inches thick; requireth its depth, so that it shall bear twice the weight of the first piece.

```
     6 deep                              Proof.
     6                                     9·7
    ---                                    9·7
    36                                    ----
     4                                    67·9
    ---                                   873
12)144                                    ----
    ---                                   94·09
    12                                     1·91 remainder added
     2 times                              ----
    ---                                   96·00
    24                                     5
    20 length                             ----
    ---                                20)480
 5)480                                    ----
    ---                                   24
    96)9·7 or 9·8, nearly for the depth
    81
    ---
187)1500
    1309
    ----
     191
```

EXAMPLE II.

Suppose a piece of timber 14 feet long, 8 inches deep, 3 inches thick; requireth the depth of another piece 18 feet long, 4 inches thick, so that the last piece shall bear five times as much weight as the first.

```
        8
        8
       ---
       64
        3
       ---
half 7)192
       -----
       27·4, &c.
          5 times
       ----
       137
          9 half the length
       ----
    4)1233
       -----
       308·25(17·5 the depth nearly
       1
      ---
   27)208(
      189
     ----
  345).1925, &c.
```

As the length of both pieces of timber is divisible by the number 2, therefore half the length of each is used instead of the whole; the answer will be the same.

PROPOSITION III.

Given the length, breadth, and depth of a piece of timber; to find the breadth of another piece whose length and depth are given, so that the last piece shall bear the same weight as the first piece or any number of times more.

RULE.

Multiply the square of the depth of the first piece into its thickness; that divided by its length, multiply the quotient by the number of times as you would have the last piece bear more than the first; that being multiplied by the length of the last piece, and divided by the square of its depth, this quotient will be the breadth required.

EXAMPLE I.

Given a piece of timber 12 feet long, 6 inches deep, 4 inches thick; and another piece 16 feet long, 8 inches deep; requireth the thickness, so that it shall bear twice as much weight as the first piece.

	Or this at full length,
6	6 depth of the first piece
6	6
36	36
4	4 thickness of the first piece
3)144	Length 12)144
48	12
2	2 by the number of times stronger
96	24
4	16 length of the last piece
8)384	144
8)48	24
6 thickness	8)384
	8) 48
	6 thickness

EXAMPLE II.

Given a piece of timber 12 feet long, 5 inches deep, 3 inches thick; and another piece 14 feet long, 6 inches deep; requireth the thickness, so that the last piece may bear four times as much weight as the first piece.

```
      5
      5
    ---
     25
      3
  -----
  12)75
 ------
   6·25
      4
  -----
   25·0
   14
  -----
   100
   25
  -----
  6)350
  -----
 6)58·333
 --------
   9·722
```

PROPOSITION IV.

If the stress does not lie in the middle of the timber, but nearer to one end than the other, the strength in the middle will be to the strength in any other part of the timber, as 1 divided by the square of half the length is to 1 divided by the rectangle of the two segments, which are parted by the weight.

EXAMPLE I.

Suppose a piece of timber 20 feet long, the depth and width are immaterial; suppose the stress or weight to lie five feet distant from one of its ends, consequently from the other end 15 feet, then the above proportion will be $\frac{1}{10\times10} = \frac{1}{100} : \frac{1}{5\times15} = \frac{1}{75}$ as the strength at five feet from the end is to the strength at the middle, or ten feet, or as $\frac{100}{100} = 1 : \frac{100}{75} = 1\frac{1}{3}$.

Hence it appears that a piece of timber 20 feet long is one-third stronger at 5 feet distance from the bearing, than it is in the middle, which is 10 feet, when cut in the above proportion.

EXAMPLE II.

Suppose a piece of timber 30 feet long; let the weight be applied 4 feet distant from one end, or more properly from the place where it takes its bearing, then from the other end it will be 26 feet, and the middle is 15 feet; then, $\frac{1}{15 \times 15} = \frac{1}{225} : \frac{1}{4 \times 26} = \frac{1}{104}$, or as $\frac{225}{225} = 1 : \frac{245}{104} = 2\frac{17}{104}$, or nearly $2\frac{1}{6}$.

Hence it appears that a piece of timber 30 feet long will bear double the weight, and one-sixth more, at four feet distance from one end, than it will do in the middle, which is 15 feet distant..

EXAMPLE III.

Allowing that 266 pounds will break a beam 26 inches long, requireth the weight that will break the same beam when it lies at 5 inches from either end; then the distance to the other end is 21 inches; $21 \times 5 = 105$, the half of 26 inches is 13 $\therefore$ $13 \times 13 = 169$; therefore the strength at the middle of the piece is to the strength at five inches from the end, as $\frac{169}{169} :: \frac{169}{105}$ or as $1 : \frac{169}{105}$ the proportion is stated thus:

$$1 : \frac{169}{105} :: \overset{\text{lb.}}{266} : \text{to the weight required}$$

```
    169
  -----
   2394
  1596
  266
  -----
105)44954(428
    420
    ---
     295
     210
     ---
      854
      840
      ---
       14
```

From this calculation it appears, that rather more than 428 pounds will break the beam at 5 inches distance from one of its ends, if 266 pounds will break the same beam in the middle.

By similar propositions the scantlings of any timber may be computed, so that they shall sustain any given weight; for if the weight one piece will sustain be known, with its dimensions, the weight that another piece will sustain, of any given dimensions, may also be computed. The reader must observe, that although the foregoing rules are mathematically true, yet it is impossible to account for knots, cross-grained wood, &c., such pieces being not so strong as those which are straight in the grain; and if care is not taken in choosing the timber for a building, so that the grain of the timbers run nearly equal to one another, all rules which can be laid down will be baffled, and consequently all rules for just proportion will be useless in respect to its strength. It will be impossible, however, to estimate the strength of timber fit for any building, or to have any true knowledge of its proportions, without some rule; as without a rule everything must be done by mere conjecture.

Timber is much weakened by its own weight, except it stand perpendicular, which will be shown in the following problems; if a mortice is to be cut in the side of a piece of timber, it will be much less weakened when cut near the top, than it will be if cut at the bottom, provided the tenon is driven hard in to fill up the mortice.

The bending of timber will be nearly in proportion to the weight that is laid on it; no beam ought to be trusted for any long time with above one-third or one-fourth part of the weight it will absolutely carry; for experiment proves, that a far less weight will break a piece of timber when hung to it for any considerable time, than what is sufficient to break it when first applied.

PROBLEM I.

Having the length and weight of a beam that can just support a given weight, to find the length of another beam of the same scantling that shall just break with its own weight.

Let l = the length of the first beam;
L = the length of the second;
a = the weight of the first beam;
w = the additional weight that will break it.

And because the weights that will break beams of the same scantling are reciprocally as their lengths,

therefore $\frac{1}{l} : \frac{1}{L} :: w + \frac{a}{2} : \frac{w + \frac{a}{2}}{L} l = W =$ the weight that will break the greater beam; because $w + \frac{a}{2}$ is the whole weight that will break the lesser beam.

But the weights of beams of the same scantling are to one another as their lengths:

Whence, $l : L :: \frac{a}{2} : \frac{L a}{2 l} = W$ half the weight of the greater beam.

Now the beam cannot break by its own weight, unless the weight of the beam be equal to the weight that will break it:

$$\text{Wherefore, } \frac{L\,a}{2\,l} = \frac{w + \frac{a}{2}}{L}\,l = \frac{2\,w + a}{2\,L}\,l$$

$$L^2\,a = 2\,w + a \times l^2$$

$\therefore a : 2\,w + a :: l^2 : L^2$, consequently $\sqrt{L^2} = L =$ the length of the beam that can just sustain its own weight.

PROBLEM II.

Having the weight of a beam that can just support a given weight in the middle, to find the depth of another beam similar to the former, so that it shall just support its own weight.

Let $d =$ the depth of the first beam;
$x =$ the depth of the second;
$a =$ the weight of the first beam;
$w =$ the additional weight that will break the first beam.

then will $w + \frac{a}{2}$ or $\frac{2\,w+a}{2} =$ the whole weight that will break the lesser beam.

And because the weights that will break similar beams are as the squares of their lengths,

$$\therefore d^2 : x^2 :: \frac{2\,w+a}{2} : \frac{2\,x^2 \times w\,x^2 + a}{2\,d} = W$$

the weights of similar beams are as the cubes of their corresponding sides:

$$\text{Hence } d^3 : x^3 :: \frac{a}{2} : \frac{a\,x^3}{2\,d^3} = W$$

$$\therefore \frac{a\,x}{2\,d^3} = \frac{2\,x^2\,w + x^2\,a}{2\,d^2}$$

$$a\,x = 2\,w + a \times d$$

$\therefore a : a + 2\,w :: d : x =$ the depth required.

As the weight of the lesser beam is to the weight of the lesser beam together with the additional weight, so is the depth of the lesser beam to the depth of the greater beam.

Note. Any other corresponding sides will answer the same purpose, for they are all proportioned to one another.

EXAMPLE.

Suppose a beam whose weight is one pound, and its length 10 feet, to carry a weight of 399.5 pounds, requireth the length of a beam similar to the former, of the same matter, so that it shall break with its own weight.

here $a = 1$
and $w = 399.5$
then $a + 2w = 800 = 1 + 2 \times 399.5$
$d = 10$

Then by the last problem it will be

$$1 : 800 :: 10$$
$$\frac{10}{8000} = x$$ for the length of a beam that will break by its own weight.

PROBLEM III.

The weight and length of a piece of timber being given, and the additional weight that will break it, to find the length of piece of timber similar to the former, so that this last piece of timber shall be the strongest possible:

Put $l =$ the length of the piece given;
$w =$ half its weight;
$W =$ the weight that will break it;
$x =$ the length required.

Then, because the weights that will break similar pieces of timber are in proportion to the square of their lengths,

$$\therefore l^2 : x^2 :: W + w : \frac{Wx^2 + wx}{2} = \text{the whole weight that breaks the beam;}$$

and because the weights of similar beams are as the cubes of their lengths, or any other corresponding sides,

$$\text{then } l^3 : x^3 :: w : \frac{wx}{l^3} \text{ the weight of the beam;}$$

$$\text{consequently } \frac{Wx^2 + wx^2}{l^2} \frac{wx^3}{l^3} \text{ is the weight that breaks the beam} = a \text{ maximum;}$$

therefore its fluxion is nothing.

$$\text{that is, } 2\,Wx\dot{x} + 2\,wx\dot{x} \frac{3\,wx^2\dot{x}}{l} = \text{nothing.}$$

$$2\,W + 2\,w = \frac{3\,wx}{l}$$

$$\text{therefore, } x = l \times \frac{{}_2W = + {}_2w}{3\,w}$$

Hence it appears from the foregoing problems, that large timber is weakened in a much greater proportion than small timber, even in similar pieces; therefore a proper allowance must be made for the weight of the pieces, as I shall here show by an example.

Suppose a beam 12 feet long, and a foot square, whose weight is three hundred weight, to be capable of supporting 20 hundred weight, what weight will a beam 20 feet long, 15 inches deep, and 12 thick, be able to support?

12 inches square
12
144
12
12)1728
144

15
15
75
15
225
12
2.0)270.0
135

But the weights of both beams are as their solid contents:

therefore 12 inches square
12
144
144 inches = 12 feet long
576
576
144
20736 solid contents of the 1st beam
20736:43200::3
3
cwt. lb.
20736)129600(6 .. 28 = the weight of the 2d beam
124416
. . 5184
112
10368
5184
5184
20736)580608(28
41472
165888
165888
. . . . . .

15 deep
12 wide
180
240 length in inches
7200
360
43200 solid contents of the 2d beam

144::135::21.5 by prop. 1.
21·5
67·5
135
270
12)2902·5
12)241·875
20·15625
112
31250
15625
15625
17·50000
16
30
5
8.0

21 cwt. 56 lbs. is the weight that will break the first beam, and 20 cwt. 17 lbs. 8 oz. the weight that will break the second beam; deduct out of these half their own weight.

20::17::8
3::14::0 half

17...3..8

Now 20 cwt. is the additional weight that will break the first beam; and 17 cwt. 3 lbs. 8 oz. the weight that will break the second: in which the reader will observe, that 10 :: 3 :: 8 has a much less proportion to 20, than 20 cwt. 17 lbs. 8 oz. has to 21 :: 56. From these examples the reader may see that a proper allowance ought to be made for all horizontal beams; that is, half the weights of beams ought to be deducted out of the whole weight that they will carry, and that will give the weight that each piece will bear.

If several pieces of timber of the same scantling and length are applied one above another, and supported by props at each end, they will be no stronger than if they were laid side by side; or this, which is the same thing, the pieces that are applied one above another are no stronger than one single piece whose width is the width of the several pieces collected into one, and its depth the depth of one of the pieces; it is therefore useless to cut a piece of timber lengthways, and apply the pieces so cut one above another, for these pieces are not so strong as before, even if bolted.

EXAMPLE.

Suppose a girder 16 inches deep, 12 inches thick, the length is immaterial, and let the depth be cut lengthways in two equal pieces; then will each piece be 8 inches deep, and 12 inches thick. Now, according to the rule of proportioning timber, the square of 16 inches, that is, the depth before it was cut, is 256, and the square of 8 inches is 64; but twice 64 is only 128, therefore it appears that the two pieces applied one above another is but half the strength of the solid piece, because 256 is double 128.

If a girder be cut lengthways in a perpendicular direction, the ends turned contrary, and then bolted together, it will be but very little stronger than before it was cut; for although the ends being turned give to the girder an equal strength throughout, yet wherever a bolt is, there it will be weaker, and it is very doubtful whether the girder will be any stronger for this process of sawing and bolting; and I say this from experience.

If there are two pieces of timber of an equal scantling (Pl. 51, *Fig. B*), the one lying horizontal, and the other inclined, the horizontal piece being supported at the points *e* and *f*, and the inclined piece at *c* and *d*, perpendicularly over *e* and *f*, according to the principles of mechanics, these pieces will be equally strong. But to reason a little on this matter, let it be considered, that although the inclined piece *D* is longer, yet the weight has less effect upon it when placed in the middle, than the weight at *h* has upon the horizontal piece *C*, the weights being the same; it is therefore reasonable to conclude, that in these positions the one will bear equal to the other.

The foregoing rules will be found of excellent use when timber is wanted to support a great weight; for, by knowing the superincumbent weight, the strength may be computed to a great degree of exactness, so that it shall be able to support the weight required. The consequence is as bad when there is too much timber, as when there is too little, for nothing is more requisite than a just proportion throughout the whole building, so that the strength of every part shall always be in proportion to the stress; for when there is more strength given to some pieces than others, it encumbers the building, and consequently the foundations are less capable of supporting the superstructure.

No judicious person, who has the care of constructing buildings, should rely on tables of scantlings, such as are commonly in books; for example, in story posts the scantlings, according to several authors, are as follows:

For 9 feet high	6 inches square.	
12 ——	8	
15 ——	10	
18 ——	12	

Now, according to this table, the scantlings are increased in proportion to the height; but there is no propriety in this, for each of these will bear weight in proportion to the numbers 9, 16, 25, and 36, that is, in proportion to the square of their heights, 36 being 4 times 9; therefore the piece that is 18 feet long, will bear four times as much weight as that piece which is nine feet long; but the 9 feet piece may have a much greater weight to carry than an 18 feet piece, suppose double: in this case it must be near 12 inches square instead of 6. The same is also to be observed in breast-summers, and in floors where they are wanted to support a great weight; but in common buildings, where there are only customary weights to support, the common tables for floors will be near enough for practice.

To conclude the subject; it may be proper to notice the following observations which several authors have judiciously made, viz. that in all timber there is moisture, wherefore all bearing timber ought to have a moderate camber, or roundness on the upper side, for till that moisture is dried out the timber will swag with its own weight.

But then observe, that it is best to truss girders when they are fresh sawn out, for by their drying and shrinking, the trusses become more and more tight.

That all beams or ties be cut, or in framing forced to a roundness, such as an inch in twenty feet in length, and that principal rafters also be cut or forced in framing, as before; because all joists, though ever so well framed, by the shrinking of the timber and weight of the covering, will swag, sometimes so much as not only to be visible, but to offend the eye: by this precaution the truss will always appear well.

Likewise observe, that all case bays either in floors or roofs do not exceed twelve feet if possible, that is, do not let your joints in floors exceed twelve feet, nor your purlines in roofs, &c., but rather let their bearing be eight, nine, or ten feet: this should be regarded in forming the plan.

Also in bridging floors, do not place your binding or strong joists above three, four, or five feet apart, and that your bridging or common joists are not above ten or twelve inches apart, that is, between one joist and another.

Also, in fitting down tie beams upon the wall plates, never make your cocking too large, nor yet too near the outside of the wall plate, for the grain of the wood being cut across in the tie beam, the piece that remains upon its end will be apt to split off, but keeping it near the inside will tend to secure it. See Plate 33, at the bottom, where the dimensions are figured.

Likewise observe, never to make double tenons for bearing uses, such as binding joists, common joists, or purlines; for, in the first place, it very much weakens whatever you frame it into, and in the second place it is a rarity to have a draught to both tenons, that is, to draw both joints close; for the pin in passing through both tenons, if there is a draught in each, will bend so much, that unless it be as tough as wire, it must needs break in driving, and consequently do more hurt than good.

Roofs will be much stronger if the purlines are notched above the principal rafters, than if they are framed into the side of the principals; for by this means, when any weight is applied in the middle of the purline, it cannot bend, being confined by the other rafters; and if it do, the sides of the other rafters must needs bend along with it, consequently it has the strength of all the other rafters sideways added to it.

DESIGNS OF ROOFS.

PLATE 39.

A and *B* show the method of trussing girders as is used by the greatest masters at this time.

C is a horizontal section of *B*.

D is a section of the butment, by cutting across *a b* in *A*.

E and *F* show the two sides of the king-bolt, at *c* in *A*, which is made with a wedge-way upon the top, so that it may force out the trusses upon the butments.

The cores of the trusses ought not to be let close into the grooves of the side beams, but should be well secured at the ends and in the middle; for suppose a weight to be laid upon the girder sufficient to bend it, if the braces are tightly fitted they will bend along with the beams, and consequently be of no use; but if they are not fitted close, the girder can never have a curvature to any sensible degree, for the side beams cannot bend without shortening their length, but in the act of bending, the braces force upon the ends, and consequently act in opposition to the side beam.

In tightening the truss work the head of your king-bolt ought to be greased, so that it may slide freely past the ends of the trusses; proceed to turn the nut of the king-bolt, and another person to hit the head at *c*, with a mallet, which will make it start every time it is hit, and give fresh ease at every turning of the nut, so that you may camber the girder to any degree that you shall have occasion for, but generally not above an inch in twenty feet.

Note. The sections *D*, *E*, and *F*, are to one-eighth part of the real size.

Plate 39.

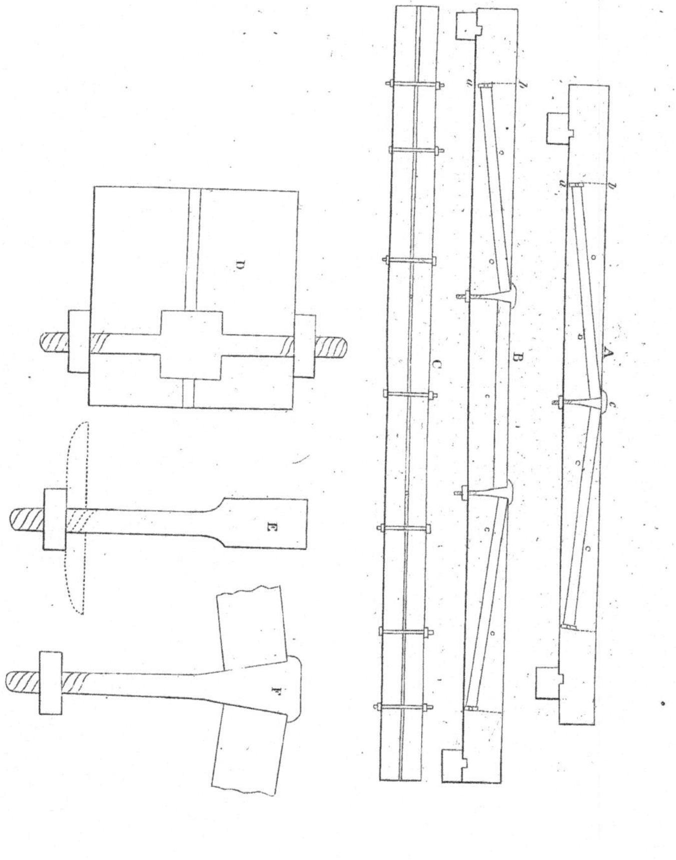

Plate 10.

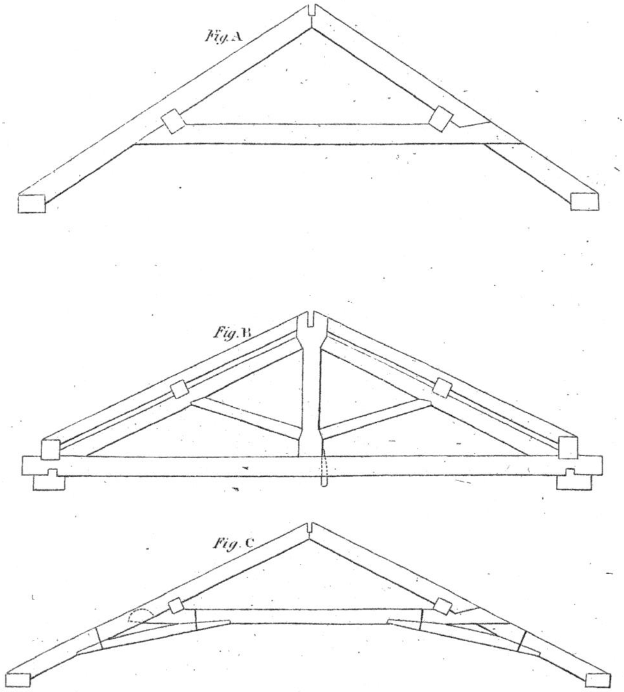

PLATE 40.

This contains the most simple construction of roofs.

FIG. *A* is calculated for a small building; at one end of the collar beam is the Carpenter's boast, what they term a dove-tail tenon; but I think rather a rule-joint, as it is worked out to a centre. This roof will do for an extent of 20 or 25 feet.

FIG. *B* is a truss for a roof, the purlines to be notched upon the principal rafters, as will be described in the following plates of ledgment roofs; this roof may be well calculated for an extent of 30 or 35 feet, the height one-fourth of the span for slate covering.

FIG. *C* is a simple construction of a roof, for the segment finish of a dome, which will answer to any of the above extents.

PLATE 41.

FIG. *A* is a roof for the purlines to be framed in, and the common rafters to come fair with the principals.

FIG. *B* is a roof calculated for a greater extent than any of the foregoing roofs, and may well extend 50 or 60 feet. Here likewise is shown the connection of the roof in the walls.

FIG. *C* is a roof supported by two queen posts, instead of a king post, to give room for a passage or any other conveniency in the roof.

Plate 41.

Fig. A.

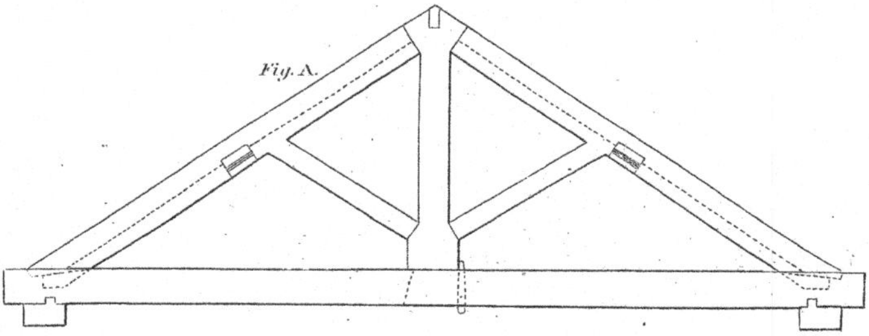

Fig. B.

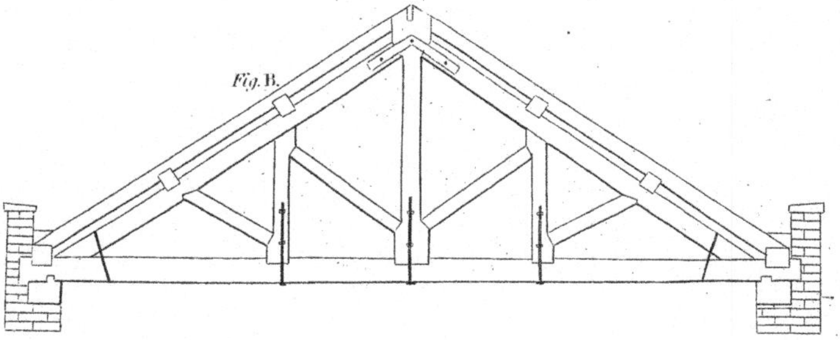

Fig. C.

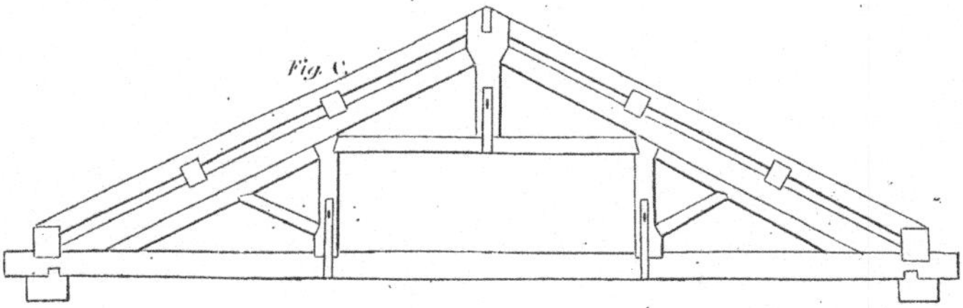

Plate 12.

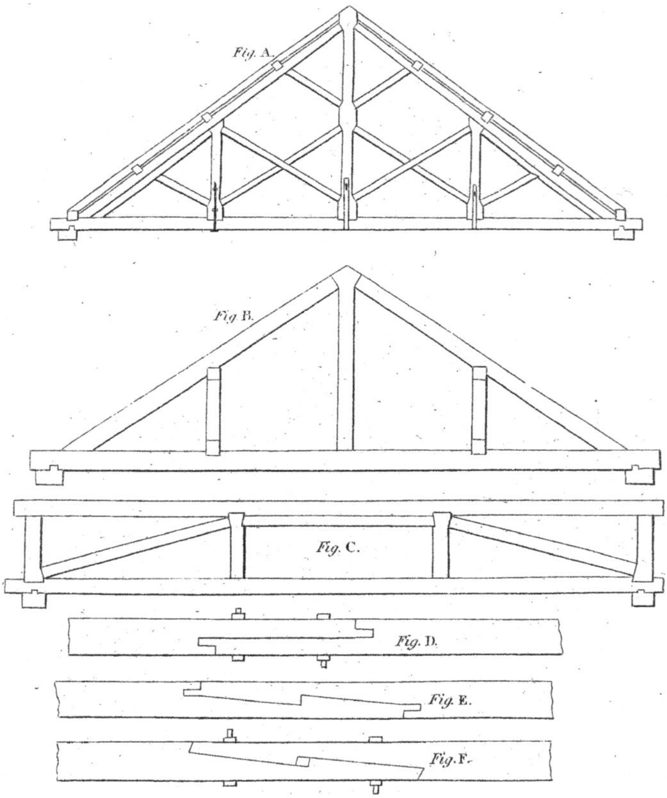

PLATE 42.

This roof is calculated for a span of seventy or eighty feet.

You will observe in this, and the foregoing roofs, that the trusses are the same in number as the purlines which they have to support; for how absurd it is to give a roof more strength than necessary! but, on the other hand, the consequences will be dangerous if too weak.

Fig. *A* is a design of a roof for a theatre, which may extend from 80 to 90 feet.

As it happens frequently in building that walls run across the roof, in such cases there will be little occasion for trussing the roof; then the purlines may be trussed, which will save one or two pair of principals, which is a considerable advantage.

Fig. *B* is a roof of this kind, which shows the ends of the purlines, and *C* shows how to truss the purline.

D, *E* and *F* are the methods of scarfing timber.

PLATE 43.

Fig. *A*. A design of a roof to finish with the parapet; when covered in the extent from 30 to 40 feet.

Fig. *B*. A roof for a theatre, the extent from 70 to 80 feet.

Fig. *C*. A design for a partition.

Plate 13.

A design of a roof to Finish with the parapet when covered in the extent from 30 to 40 feet.

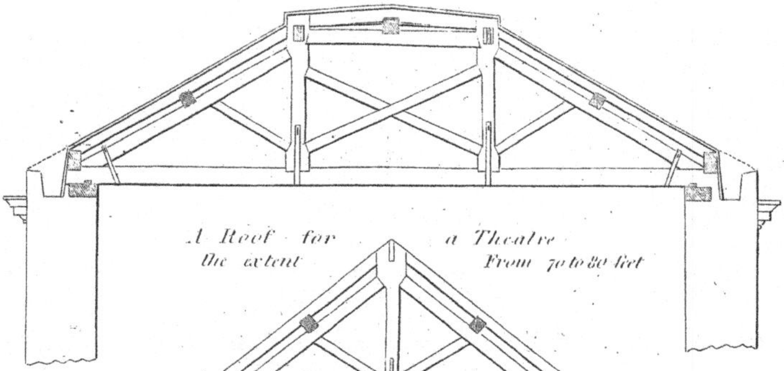

A Roof for a Theatre
the extent From 70 to 80 feet

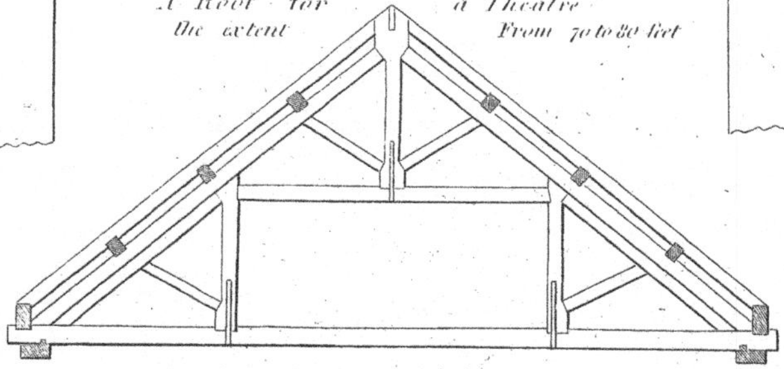

A design for a Partition

PLATE 43.

FIG. *A*. A design of a roof to finish with the parapet, when covered in the extent from 30 to 40 feet.

FIG. *B*. A roof for a theatre, the extent from 70 to 80 feet.

FIG. *C*. A design for a partition.

Plate 13.

A design of a roof to Finish with the parapet when covered in the extent from 30 to 40 feet.

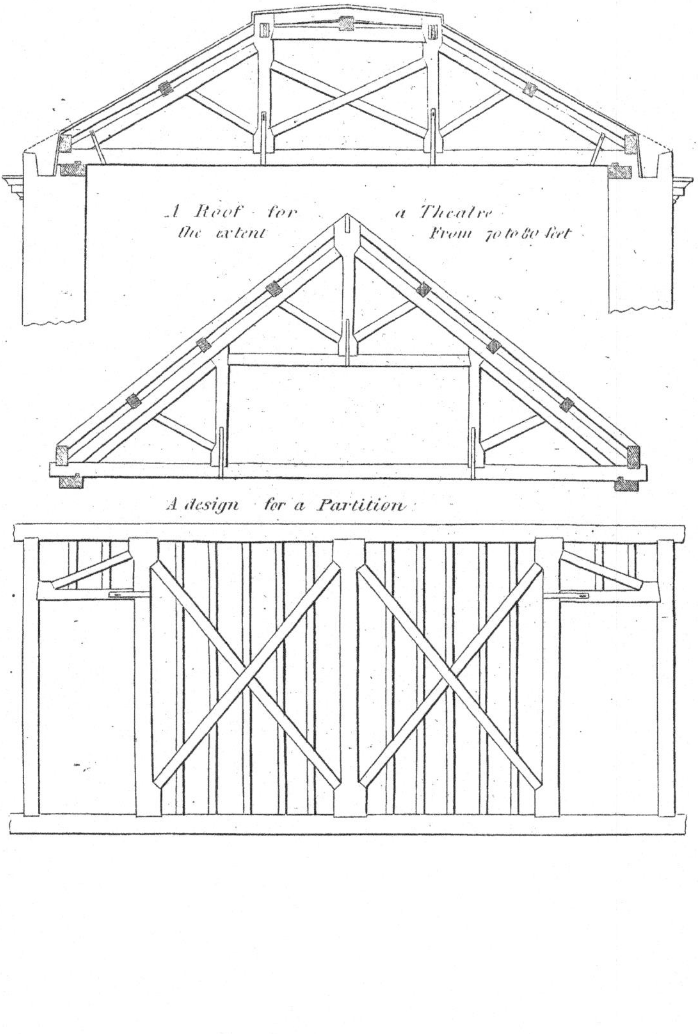

Plate 44.

Fig. A.

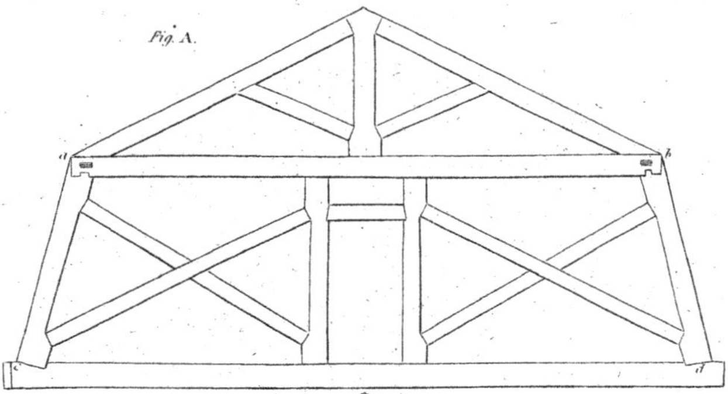

Fig. B.

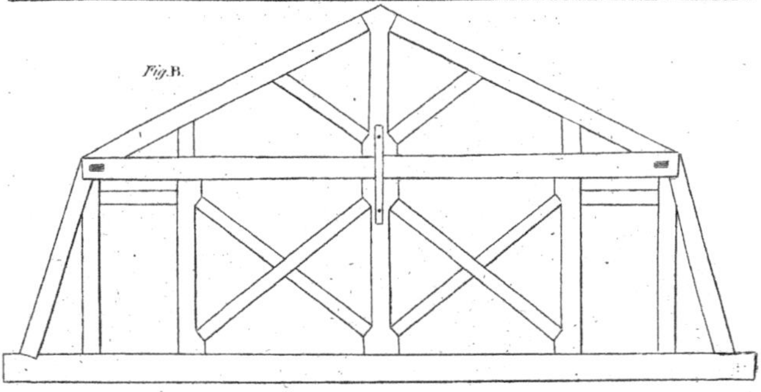

Fig. C.

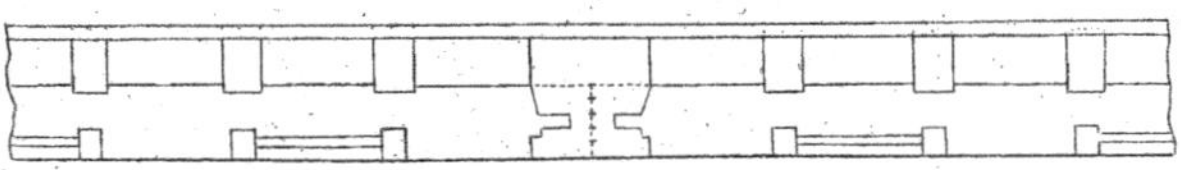

PLATE 44.

FIG. *A* is a curb roof, with a door in the middle of the partition; the beam *a b* to run quite across the roof.

FIG. *B* is a roof calculated for two rooms.

FIG. *C* shows the method of framing a bridge-floor.

PLATE 45.

FIG. *A* is the design of an M roof, which is useful in some cases where the span is great, and no wall between, and the roof is required not to appear of a great height; but this seldom happens in practice, for if there is any wall between the external walls, the roofs are in general made double, as is shown at *figures B, C,* and *D.*

Plate 15.

Fig. A.

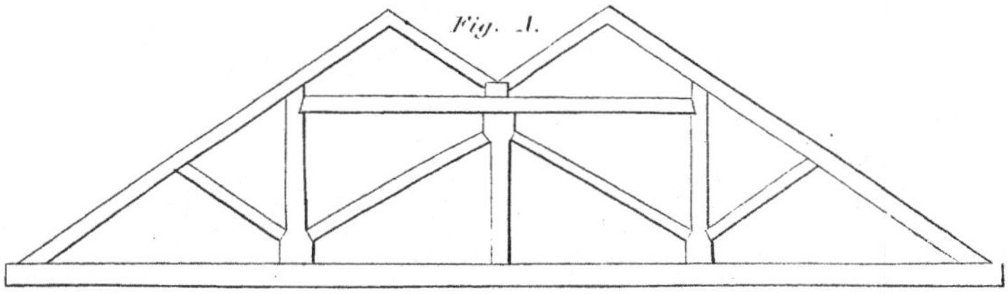

Fig. B.

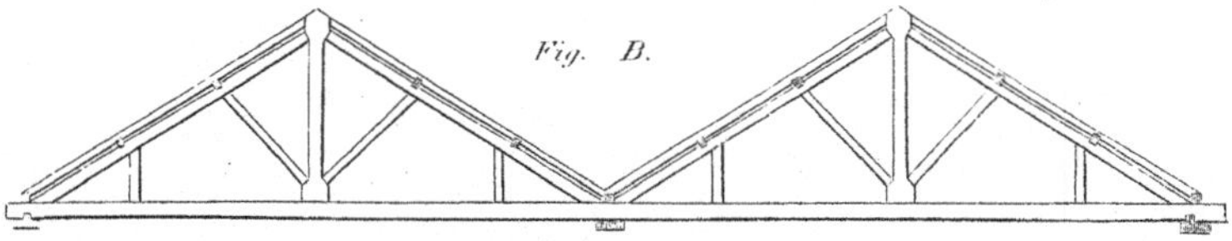

Fig. C.

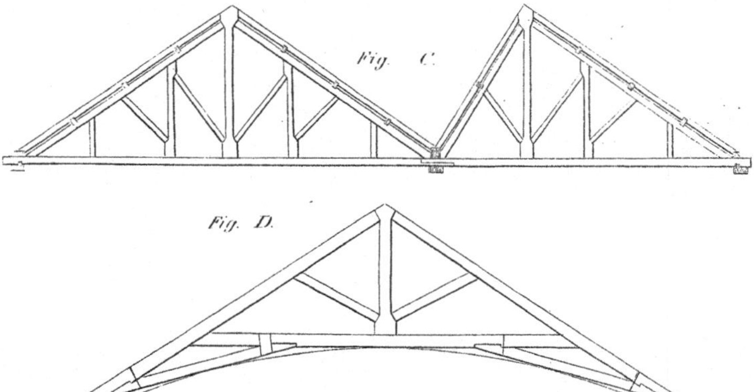

Fig. D.

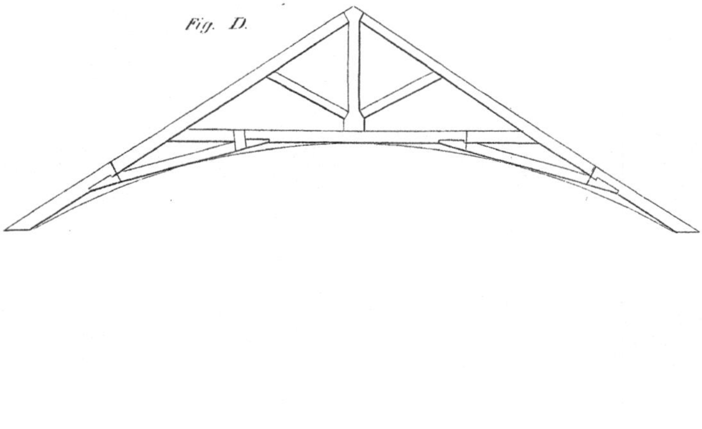

Plate 46.

Fig. A.

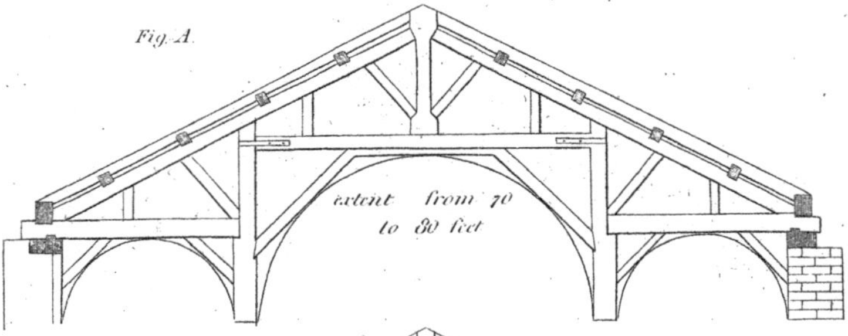

Fig. B.

Fig. C.

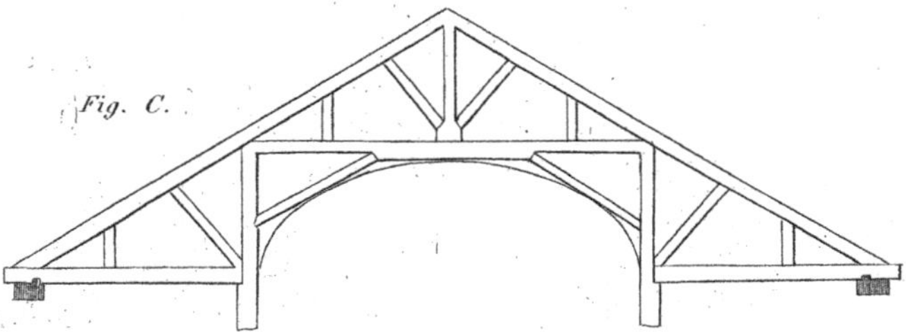

PLATE 46.

FIG. *A* is a design for a church roof, the extent marked on the plate.

FIG. *B* is a design of the same kind, but may be applied to an extent much greater. These two roofs, when finished, will be the same in every respect with the cylindro-cylindric arches described in Plate 23; as the manner there shown of fixing the ribs will not be different in this, I refer the reader to the description of that plate.

FIG. *C* is another design for a church roof, where the ceiling over the galleries is to finish level.

PLATE 47.

Fig. *A* is a design for a domical roof; *B* shows the manner of framing the curb for it to stand upon, the section of the curb being also shown upon the bottom of *fig. A*.

N. B. This design is nearly the same as that constructed for the dome of the Pantheon, London, which was burnt down.

Plate 47.

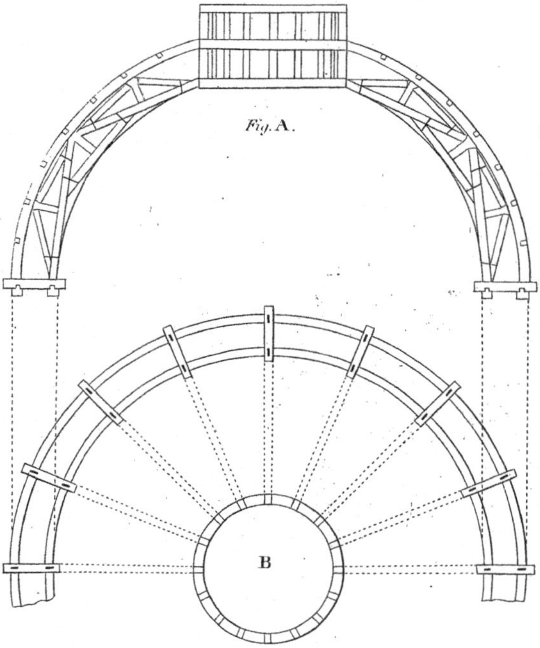

Plate. 48.

Fig. A.

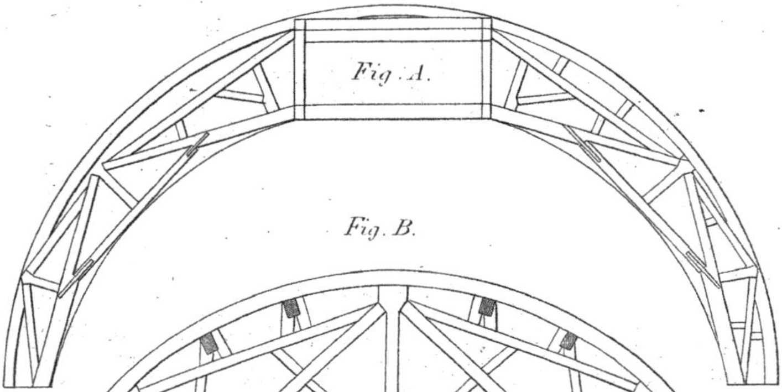

Fig. B.

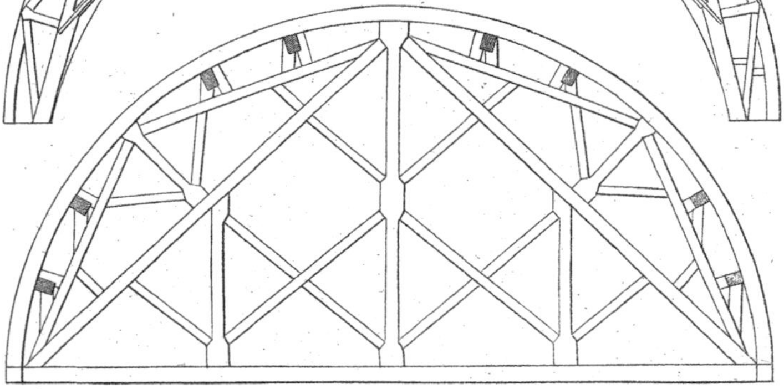

Fig. C.

PLATE 48.

FIG. *A* is another design for a domical roof; the bottom of it is made into a very narrow compass, in order to gain room within the dome.

FIGURES *B* and *C* are designs for circular and elliptical trusses for bridges, &c. These trusses may also be applied to roofs where there is no cavity wanted within.

PLATE 49.

FIG. *A* is a design for a story post and breast-summers.

FIG. *B* is a design for a bridge. *C* is a section across. *D* is part of the plan, which also shows the manner of fixing the piles. *E* shows half the plans of the bridgings.

Plate 49.

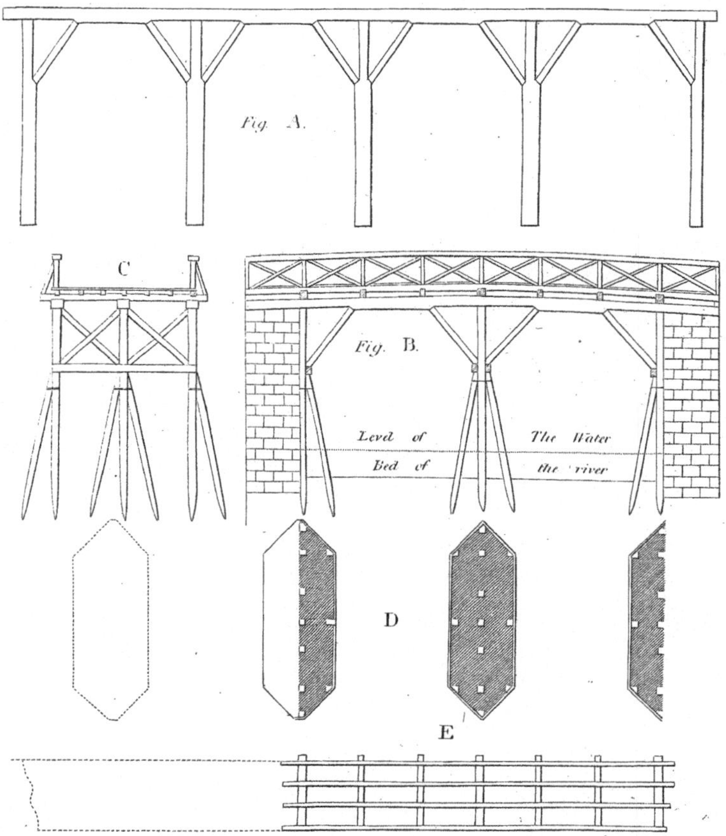

Plate 50.

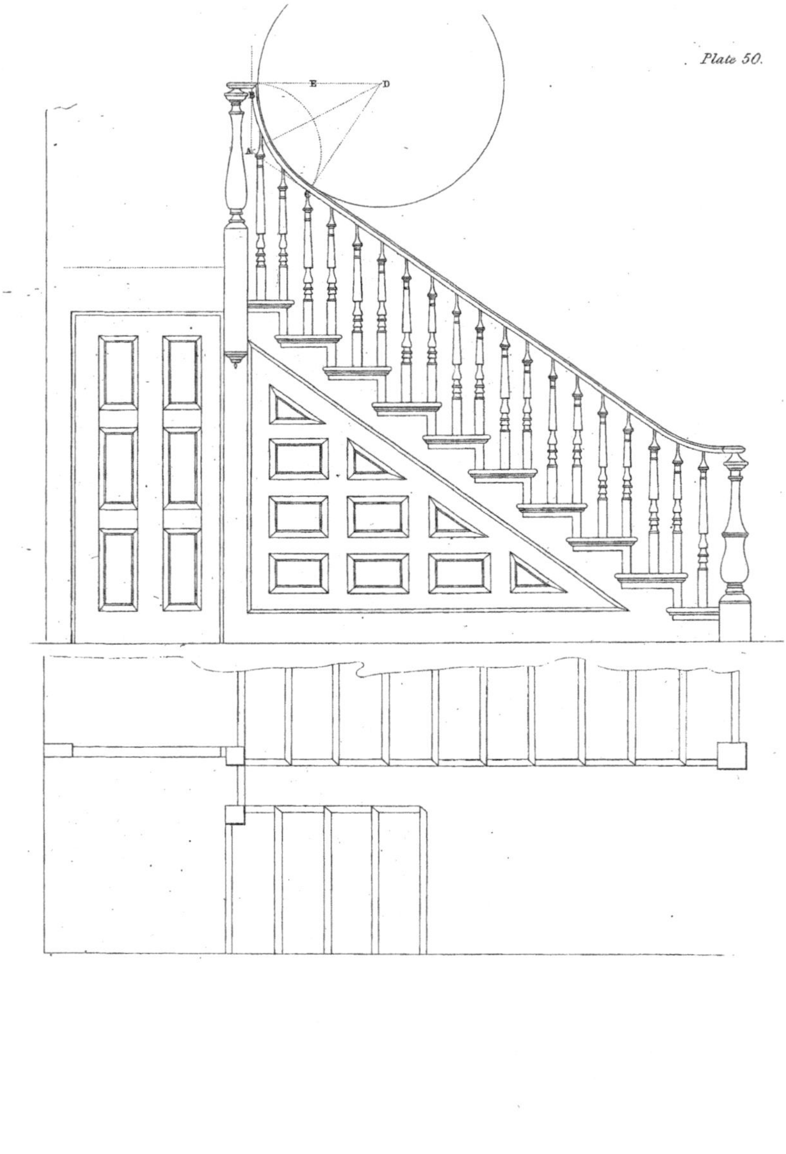

HAND-RAILING.

PREFACE.

IN that elegant branch of the building art called Joinery, Stairs and Hand-railing take precedence. For the manner of finding the face and falling moulds, I have laid down correct methods founded on the most obvious principles, and which have been put in practice by myself, and by those who attended the instructions given by me, in this art, some years since, and found to answer well in every case.

The superior advantages in every respect of the new plates on hand-railing, over those published in the former editions of this work, will, I trust, be deemed a sufficient reason for the change made by me in this department of the Carpenter's Guide. I have retained those plates on hand-railing, by P. Nicholson, which are considered useful, and hope that the alterations made in this department of his work, will meet the approbation of Carpenters generally. In conclusion, I think it proper to say, that, for the method of finding the butt joints, Plate 55, I am indebted to an eminent stair-builder of this city, whose mechanical skill in joinery, I, with others, hold in high estimation.

WILLIAM JOHNSTON,
ARCHITECT, *Philadelphia.*

PLATE 50.

Plan and elevation of a newel post stair-case. Scale ½ an inch to a foot.

To draw the Ramp.

Make *A B* equal to *A C*, draw *C D* at right angles with the rail, also produce the horizontal line *E* until it intersects at *D*, which is the centre of the ramp.

PLATE 51.

Plan of stair-case on *P*late 50, to a scale of 3 inches to a foot.

Plate 51.

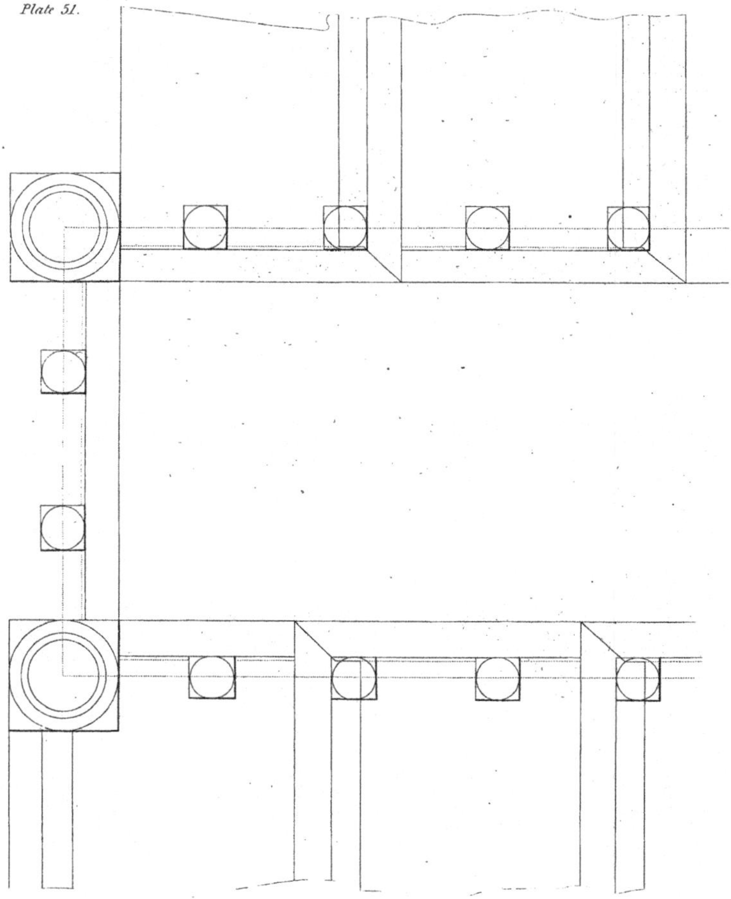

Plate 52.

PLATE 52.

Plan of a stair-case, showing how to arrange the steps at the circle, as to allow the string-board to be formed, without the easing (usually made) on its lower edge, caused by making the tread of the steps greater or less at the circle, than at the flyers; this plan also admits of the balusters being the same length, and nearly the same distance apart as at the flyers. For manner of drawing, see Plate 53.

PLATE 53.

Well Hole of Plate 52 to a larger scale.—Describe the equilateral triangle *A B C*, draw the right line *E D*, touching the face of the string board, produce *C A* and *C B*, to intersect it at *D* and *E*, then is the right line *E D* equal to the semi-circumference of the circle.

Upon this line lay off the tread of a step *O* P, and draw lines from those points to the centre *C*, which gives the position of the risers *K K*, on the circular part of the string-board, at the points *S S*. Also the position of the same riser on the carriage is found, by adding to the distance *D* P or *O E*, whatever may be required to make a full tread, and place it from *A* to *W*, which gives the position of the risers *N N*, and *K K* on the carriages.

Plate. 53.

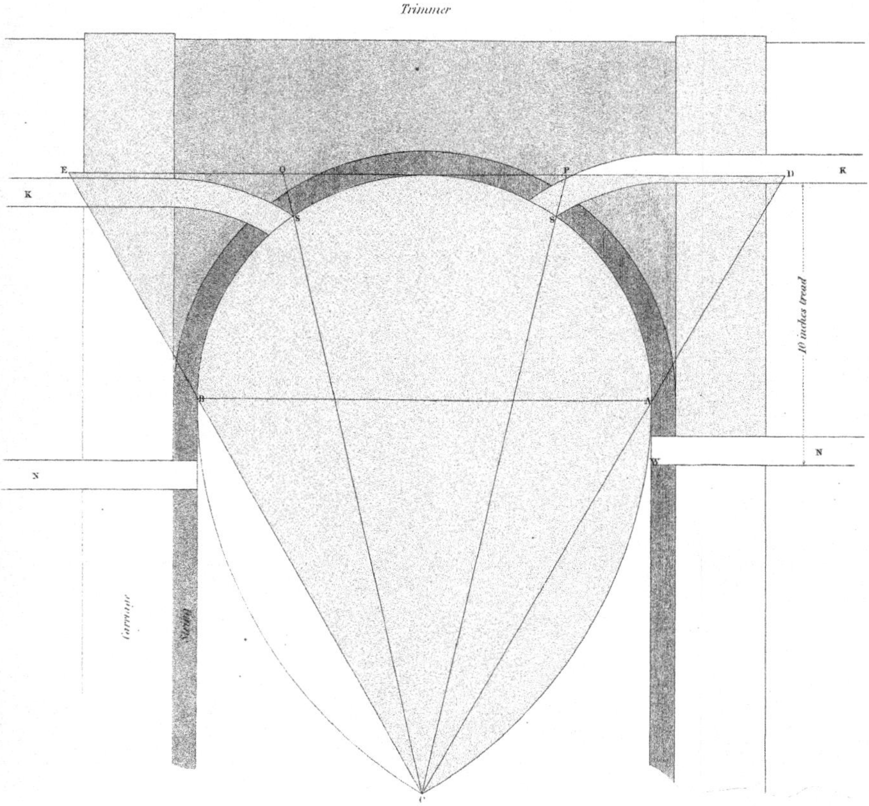

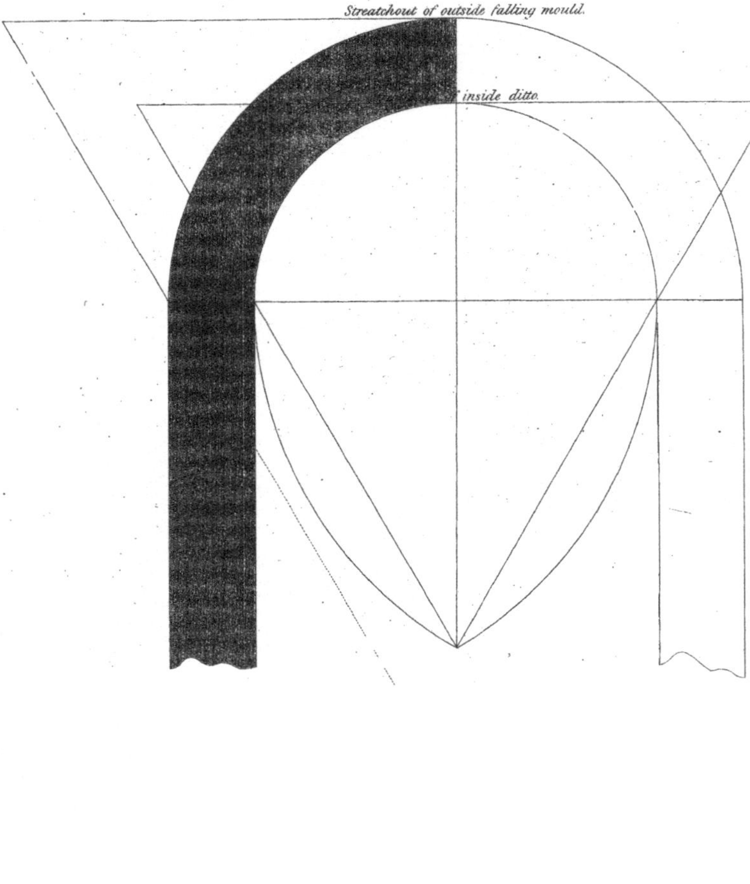
Streatchout of outside falling mould.
inside ditto.

PLATE 54.

Being a plan of the rail for Plate 53; the stretchout is found by the same method as in the foregoing plate.

PLATE 55.

To draw the falling moulds for the rail piece as shown in Plate 56. Let fig. 1 be the pitchboard, and *A* the beginning of the circular part of rail, from the line *A*, place the stretchout of the inside and outside of the rail *B B*, and from the line *C* set up the height *B F* or *B E*, which is equal to a rise and a half.—Having found the stretchout and height, place half the thickness of the rail on each side of the central points *F E* and *H*, as denoted by the small circles, connect the upper heights *F E* with the lower one *H*, which completes the falling moulds.

To cut the centre or butt joints, divide the distance *F E* into two equal parts, draw the line *H L*, and at right angles with this line draw the lines *F J* and *E K*, which gives the joint required.

Plate 55.

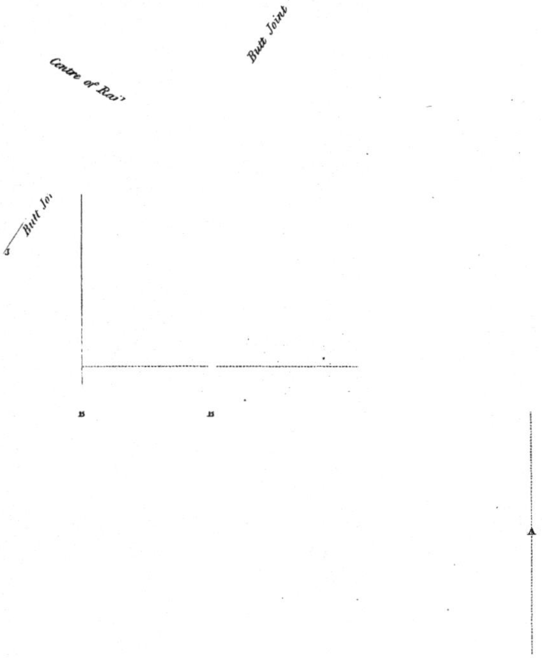

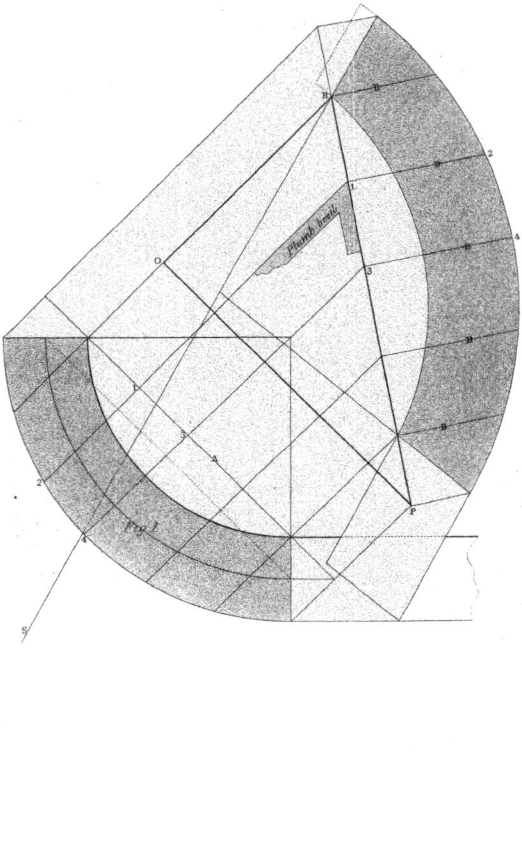
Plumb bevil
Fig. 1

PLATE 56.

To draw the face mould for plan of stairs, Plate 53. Let fig. 1 be the plan of the rail, draw the chord line *A*, also the base line *O* P, at any convenient distance above it; make the length of this line equal to the dotted line on the plan, and set the height *N E* of the falling moulds, Plate 55, from *O* to *R* draw *R P*, then draw ordinates through the plan perpendicular to the base line, touching the chord *R P*; also draw the ordinates *B B*, *B B*, &c., at right angles with this chord, and make the distances 1, 2, 3, 4, &c., on the plan equal to 1, 2, 3, 4, &c., of the face moulds.

The circular part being found, draw the straight part by keeping it parallel to the line of the joint *S*.—In applying this mould to the plank, care is to be taken that the chord *R* P be kept parallel to its edge.

As this method of finding the face mould does not admit (if 3 inch plank be used) of more than 4 or 5 inches of straight rail being attached to the circular part, another method is shown on Plate 58, by which the straight rail can be extended at pleasure.

PLATE 57.

Are the same face moulds as shown on Plate 55,—and is here reproduced to show the method of finding the spring of the plank, before drawing the face mould, Plate 58.

Let fig. 1 be the plan of the rail, draw the line *A B*, at right angles with *C D*, also draw the dotted line *D B*, which is a continuation of the greater side of the pitchboard, to intersect it at *B*—then is the distance *O* P the spring of the plank.

Plate 57.

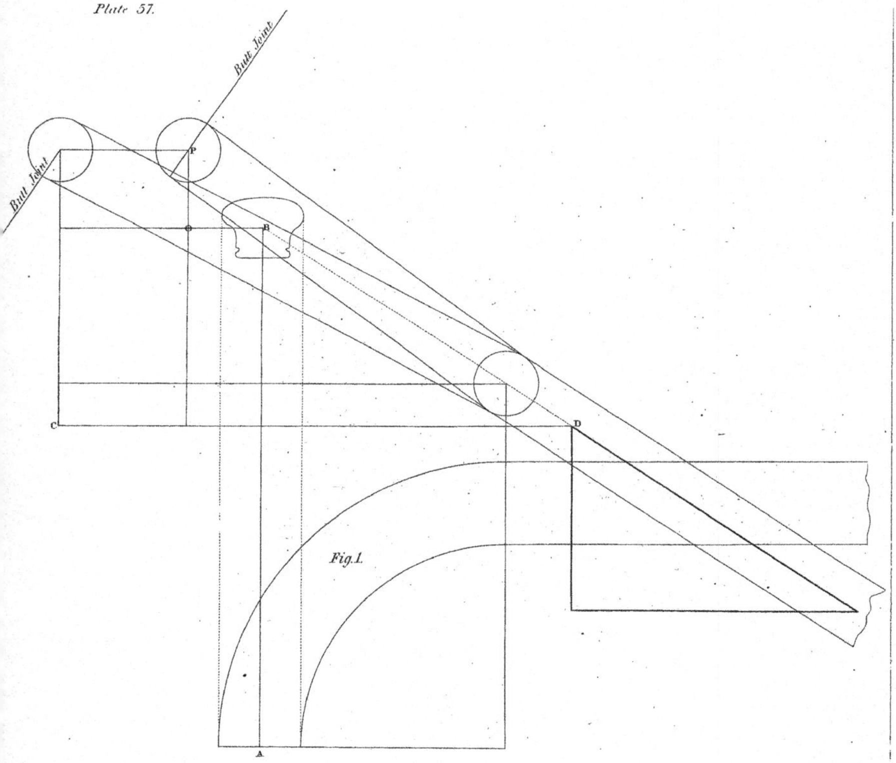

Fig.1.

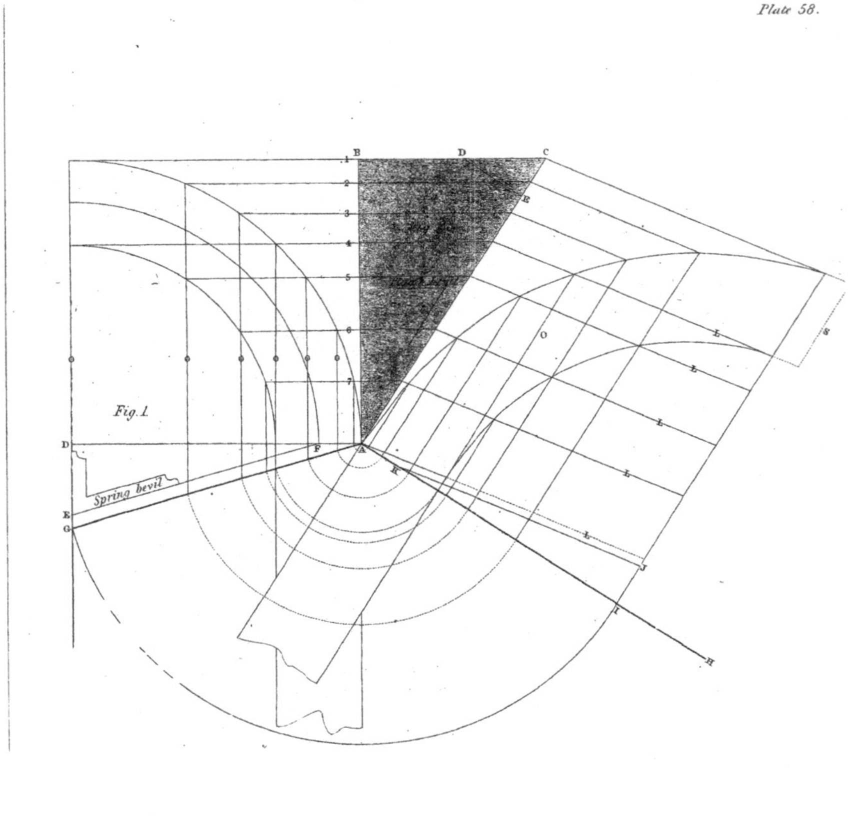
B
D
C
E
1
2
3
4
5
6
7
Fig. 1
D
F
A
K
E
G
Spring bevil
O
L
L
L
L
L
S
J
I
H

PLATE 58.

To draw the face mould *O*.

Let fig. 1 be the plan of the rail, and fig. 2, the pitchboard; draw the ordinates 1, 2, 3, 4, &c., at right angles to *A B*; place the spring *O P*, Plate 57, from *C* to *D*; draw *D E*, at right angles to *A C*. Make *D E*, fig. 2, equal to *E D*, fig. 1; draw *E F*, also *A G* parallel to it, draw the lines 0, 0, 0, &c., from the points made by the intersection of the ordinates with the convex and concave sides of the rail, and produce them to the line *A G*,—draw *A H* at right angles with *C A*, make *I J* equal to *E C*, fig. 2, join *J K*, parallel to *J K*, draw the ordinates *L L*, &c., and with *A* as a centre, draw the concentric dotted lines to intersect with the line *A H*, and continue those lines parallel to *A C* and at the points of intersection with the ordinates, trace the face mould *O*.—The dotted line *S* shows the over wood for cutting the square or butt joint.

PLATE 59.

Plan of a stair-case having 8 winders.

The manner of finding the stretchout of the convex and concave sides of the rail is the same as in the foregoing plates.

For the falling moulds and face moulds of this stairway, see Plates 60 and 61.

Plate 59.

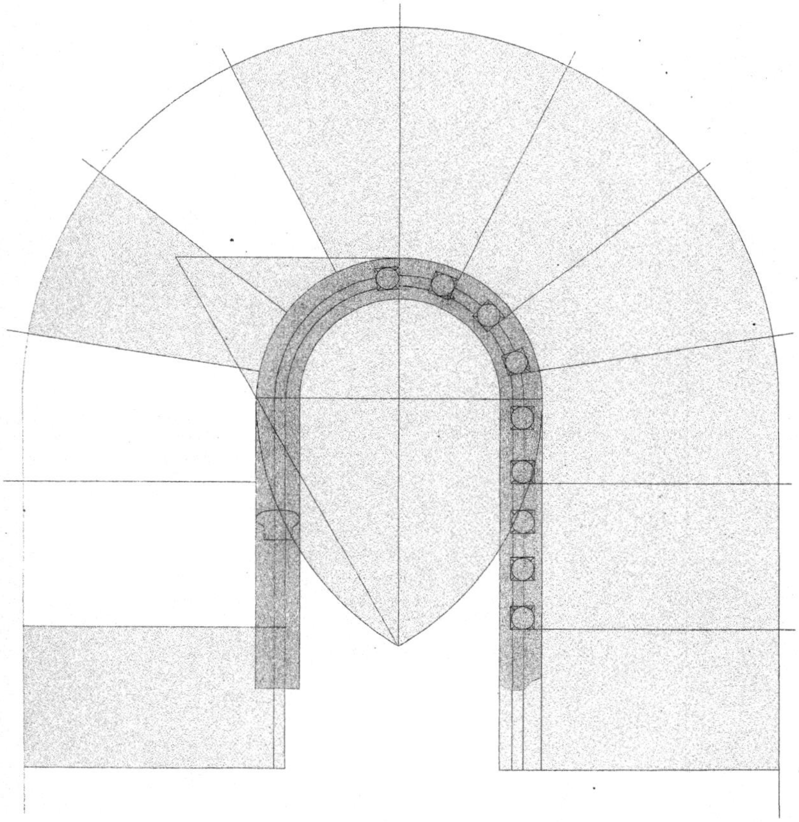

Plate 60.

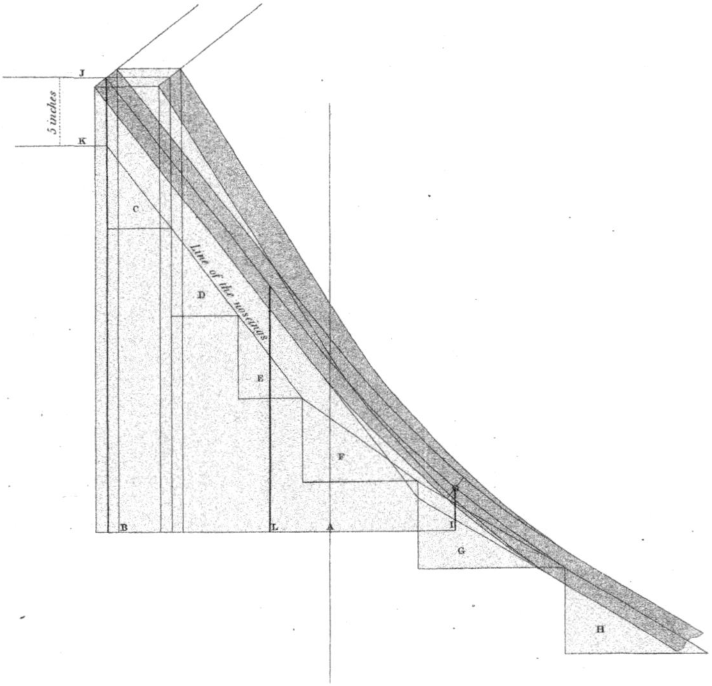

PLATE 60.

To draw the falling moulds for stair-case, Plate 59.

Let *A B* be the stretchout of the convex side of the rail, and *A I*, the straight portion attached to the circular part,—draw the flyers *H* and *G*, and the four winders *C D E* and *F*, also draw the base line *B I*, at any convenient distance below the point *O*; raise the rail 5 inches above the line of the noseings, from the point *K* to *J*, also make the distance *A L* equal to the distance *A L*, Plate 61, and proceed to draw the moulds as before described in Plates 55 or 57.

PLATE 61.

To draw the face moulds for the stair-case, Plate 59.

Let fig. 1 be the plan of the rail,—draw the chord *A B*, also the line *C D*, parallel to it, make the lines *E F G*, equal in height to the lines *B L I*, Plate 60. Join *G H*, also draw *I J* parallel to *C D*, and at the point *J* let fall the perpendicular line *J K*.—Join *K M*, which gives the directing ordinate on the plan; draw a sufficient number of ordinates to meet the chord *N O*, also draw *I P* at right angles with the line *H G*, and with the distance *K M*, from the point *J* as a centre, bisect the line *I* P at P. Join *J* P, which gives the directing ordinate for the face mould—transfer the ordinates from the plan, and set them from the chord *O N*, and through the points trace the face mould. Take the distance *I S*, set it from *Y* to *Z* on the chord, and join *Z m*; with this bevel the edge of the plank must correspond before you apply the face mould to its upper and lower surface.

Plate 61.

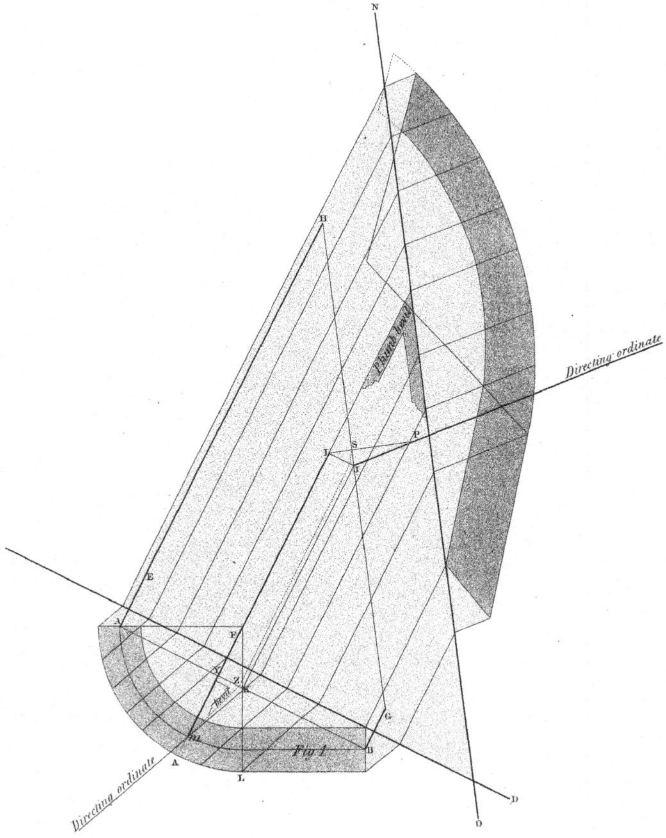
N
H
Plumb bevel
Directing ordinate
L
S
P
T
E
A
F
Y
Z
X
bevel
G
m
Fig 1
B
A
L
Directing ordinate
D
O

Plate 62.

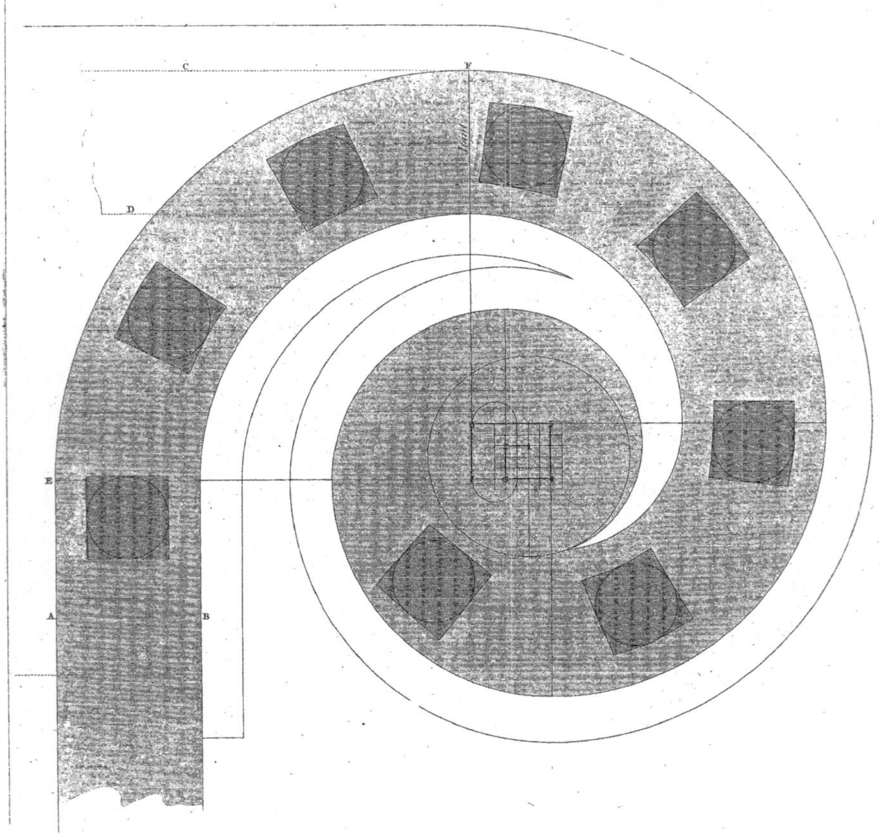

PLATE 62.

To draw the Scroll of a Hand-rail.

Make a circle $3\frac{1}{2}$ inches diameter, divide the diameter into three equal parts,—make the square in the centre equal one of these parts, and divide each of its sides into 6 equal divisions, with the centres 1, 2, 3, 4, &c., complete the outside revolution, set the width of the rail from *A* to *B*, and go the reverse way to complete the inside revolution.

If a scroll of less diameter be required—draw the dotted lines *C D* for the straight part of the rail, and a scroll of one quarter revolution less is given.

PLATE 63.

To draw the falling moulds for the scroll.

Let fig. 1 be the pitch-board, and *A B* the stretchout of the convex side of the rail *E F*, Plate 62,—lay half the thickness of the rail on each side of the upper line of the pitch-board, shown by the small circle; draw the line *D C*, and the horizontal line *F C*, and complete the under edge of the mould by intersecting lines.

In forming this easeing the mould should not come to a level at the joint, but continue to descend for a few inches past it.

Fig. 1.

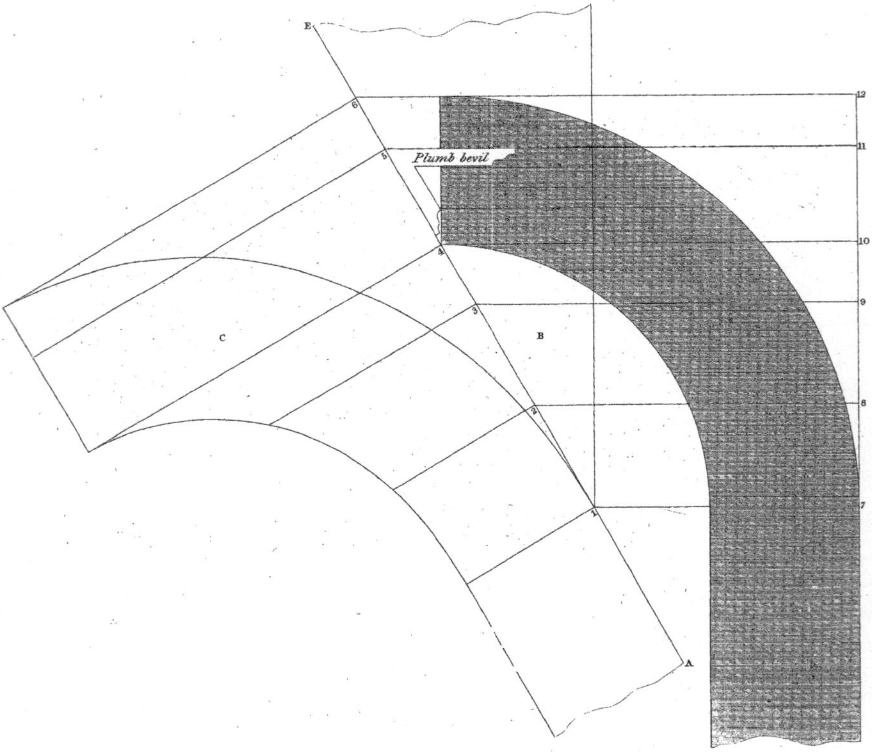
E
12
6
11
5
Plumb bevil
4
10
9
3
C
B
8
2
7
1
A

PLATE 64.

Face mould for the twist of the scroll. *B* the pitchboard.

Draw a sufficient number of ordinates through the plan, and from the points 1, 2, 3, &c., draw ordinates at right angles to the line *A E*, and transfer the distances 1, 7, 2, 8, &c., upon these lines, and trace the face mould through the points.

PLATE 65.

Plan of the curtail Riser. When the scroll is drawn, set the projection of the nosing without, and draw it equally distant from it, which will give the form of the curtail step, then set the distance which the riser is back from the line of nosings, and you will get the form of the curtail riser.

Plate 65.

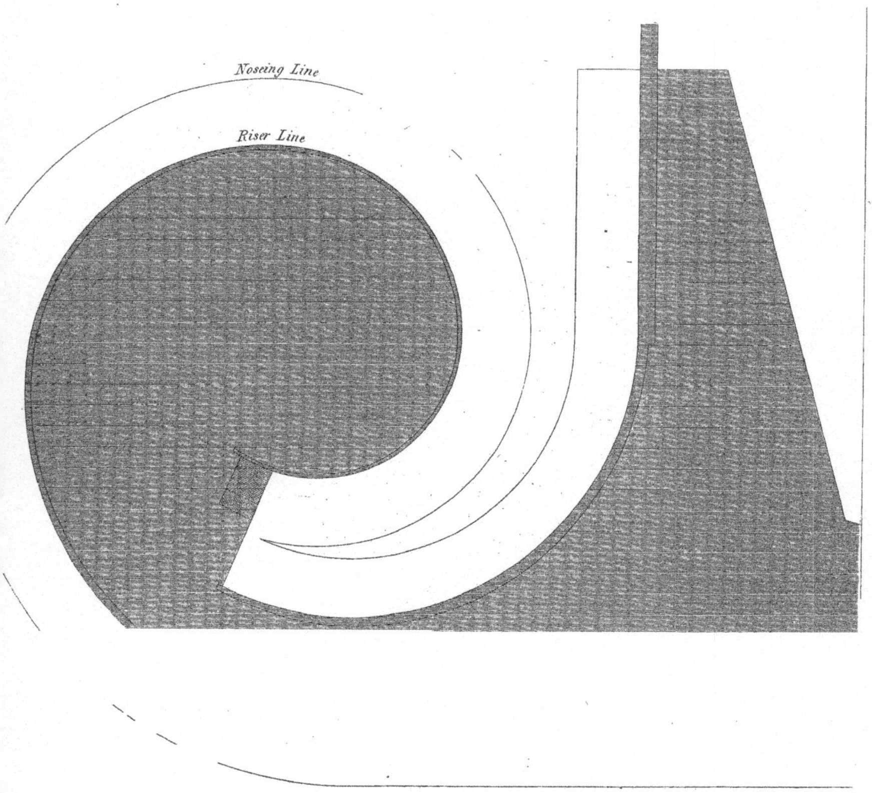

Plate 66.

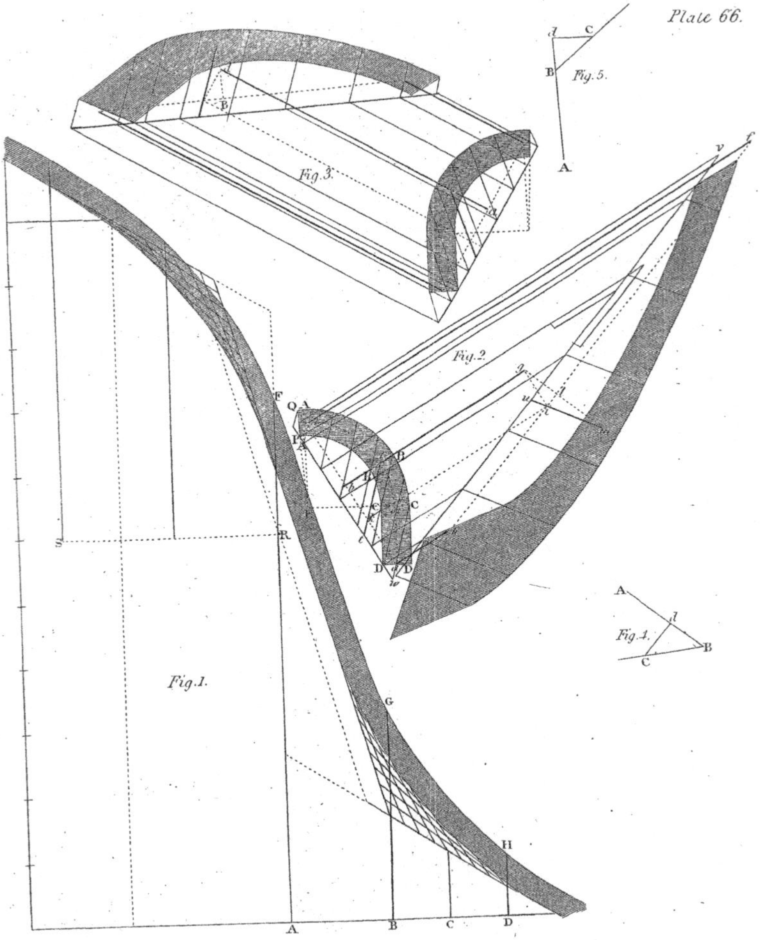

PLATE 66.

PART 1.

To find the Moulds for executing a Rail with a Semicircle of Winders.

Figure 1. The falling mould as here drawn does not follow the line of nosings, but is raised six inches, as is the practice with several handrailers, in order that the rail should not approach nearer to the nosings of the winders than to the flyers. This example is adapted to a stair with ten winders in the semi-circumference.

Fig. 2. The plan and face mould of the lower quarter winders.

Fig. 3. The plan and face mould of the upper quarter winders. *A B C* the convex side of the quadrantal part of the plan, *C D* a straight part intended to be wrought on the same piece with the twisted part. In this method, the plane of the top of the face mould is supposed to rest upon the upper extremities of three straight lines or slender rods perpendicular to the plane of the seat of the said face, and these three perpendicular lines to rise from three points, in a line dividing the breadth of the plan everywhere into two equal parts of the rail piece; and these three points to be so situated that one may be at each extremity, one at the end of the quadrant, and one at the end of the straight piece, and the intermediate point at the intersection of a perpendicular drawn from the centre of the plan of the rail piece to a straight line joining the two extreme points. Let each of the upper extremities of the three perpendicular lines be called resting points, and let the feet of the perpendiculars be called the foot of the heights of the rail piece, which will therefore be the same as the seats of the resting points; and let the three perpendicular lines themselves be called the heights of the rail piece, and their places distinguished by the lower height, the middle height, and the upper height.

For *fig.* 2, let *a* be the foot of the upper height, *b* the foot of the middle height, and *d* the foot of the lower height; join *a d*; draw *a f b′ g* and *d h* perpendicular to *a d*, the middle one *b g* being drawn from the centre *E*. Let *A, B, C, D, figure* 1, be the points corresponding to *A, B, C, D*, in the convex side of the plan, *figure* 2, and let *A F, B G* and *D H* be the heights of the rail piece, *figure* 1; make *a f b′ g, d h, figure* 2, respectively equal to *A F, B G, D H, figure* 1: join *f h, figure* 2: draw *g i* parallel to *b d*, cutting *f h* at *i*: draw *i k* parallel to *g b*, cutting *b d* at *k*: draw *g l m* perpendicular to *f h*, cutting *f h* at *l*; join *b k*; from *i*, with the distance *b k* describe an arc at *m*; join *i m*: draw *D′* P parallel to *d a* from the extremity *D* of the concave side; produce the convex quadrant *C B A* to *Q*, and the concave quadrant *C′ B′ A* to *P*, and radiate the line P *Q* from the centre *E*, which will complete the whole plan of the rail piece; the part P *A′ A Q* will make a sufficient allowance for the cutting of the joint. Draw ordinates parallel to *b k* cutting the chord *D′ P*, the concave side of the plan, and the convex side of the same: produce *b k* to meet *D′* P in *t*, and produce *m i* to *u*, making *i u* equal to *k t*: through *u* draw *v w* parallel to *f h*; from the points where the ordinates intersect *D′* P, draw lines parallel to *a f, b g*, or *d h*, cutting *v w*: from the cutting points in *v w* draw lines parallel to *u m* as ordinates: transfer the interior

PLATE 66.

PART 2.

ordinates from the plan to the face mould, also transfer the exterior ordinates of the plan to the face mould, applying them from the intersected points in the chord *v w*, and through the points thus set off, trace the concave and convex curves, as also the straight part of the mould; observe, however, that as the straight part of the mould is a parallelogram, if three points on two contiguous sides are found, joining the middle point to each of the other two, gives two sides; each of the other two remaining sides is found by drawing a line parallel to its opposite side.

In the same manner the mould *figure* 3, is to be found; but the base line of the heights is taken upon any convenient line *S R*, parallel to *A D*, *fig.* 1, so as to shorten the height lines, as otherwise *figure* 3 would occupy more space than might be found at all times convenient, and at any rate the shortening of the heights will shorten the time of drawing *figure* 3, as shorter lines can be drawn sooner than longer ones. The distance between the height lines of the upper rail piece is the same as those for the under rail piece.

To find the spring of the plank *fig.* 4: draw any straight line *A B*; from which cut off *B d* equal to *g l*, *figure* 2: draw *d C*, *figure* 4, perpendicular to *A B*: make *d C* equal to *b b'*, *figure* 2, and join *B C*, *figure* 4, then the angle *A B C* is denominated the spring of the plank, and the angle is said to be acute when the planes of the top and edge of the plank form an acute angle with each other: but when these two form an obtuse angle with each other, the spring is said to be obtuse.

To find the Spring of the Plane at an Obtuse Angle.

In *figure* 5, let *A B* be any straight line, which produce to *d*: make *B d* equal to *B d*, *figure* 3: in *figure* 5 draw *d C* perpendicular to *A B*: made *d C* equal to *a e*, *figure* 3, then *A B C* will be the spring of the plank at the obtuse angle.

Both these bevels are supposed to be applied from the top of the plank: but if the complemental angle of the obtuse spring bevel which answers to the upper wreath piece be taken, then the lower spring is applied from the top in order to give the spring of the lower wreath piece, and the complemental spring of the upper wreathed piece to the lower edge of the said piece.

The reader will perceive that in *figures* 2 and 3, though the plan is the same in both, but their position inverted, the face mould of the lower wreath piece is much longer than that of the upper one. This circumstance is owing to the middle parts of the falling mould being raised over the nosings of the winders, and the more it is raised above the winders the greater will the face mould of the lower wreathed piece exceed the length of the mould of the upper wreathed piece, and unless that (if supposing a line passing through the middle of the falling mould bisecting the breadth of the same), the distance of the line thus passing be the same over the winders as over the flyers, the two face moulds can never be equal.

Plate 67.

An elliptical Plan & section of a staircase. For its lines see the next Plate

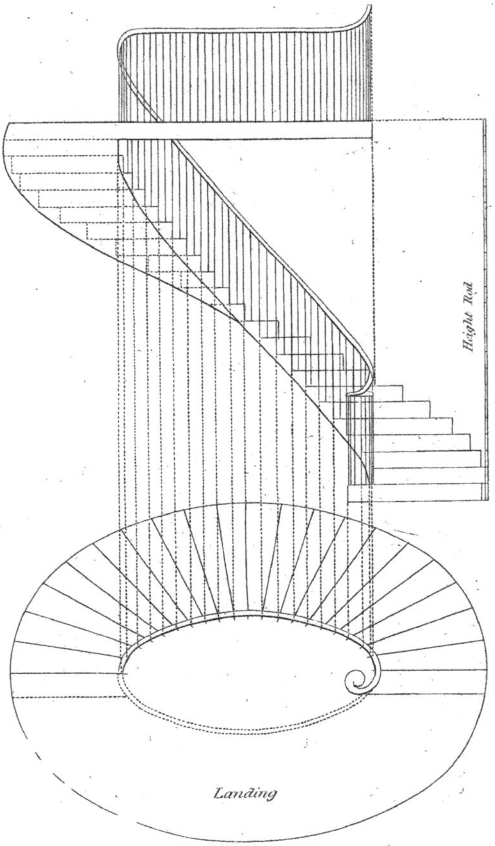

PLATE 67.

Plan and elevation of an elliptical stair-case. In every kind of stair-case whatever, the breadths of the heads of the steps are always reckoned on a line, bisecting their length, or at 18 inches distant from the rail. In this example, the steps are divided into equal parts, both at the rail and at the wall. This division will make the falling mould straight on the edges, and consequently will form an easy skirting, as well as an easy rail.

PLATE 68.

How to draw the Face Moulds of an elliptic Stair.

The plan and section being laid down as in Plate 67, the reader will observe, that the ends of the steps are equally divided at each end; that is, they are equally divided round the elliptic wall, and also at the rail. In this plate, the rail is laid down to a larger size than that in the last plate: the plan of this rail must be divided round, into as many equal parts as there are steps; then take the treads of as many steps as you please, suppose 8, and let *h h* at *fig. H*, be the tread of 8 steps from *H*; on the perpendicular *h m* set up the height of as many steps, that is 8; and draw the hypothenuse *m h*, which will give the under edge of the falling mould. The reader will observe that this falling mould will be a straight line excepting a little turn at the landing and at the scroll, where the rail must have a little bend at these places, in order to bring it level to the landing and to the scroll; then mark the plan of your rail in as many places as you would have pieces in your rail (in this plan are three); then draw a chord line for each piece to the joints; also draw lines parallel to the chords, to touch the convex side of the plan of the rail; from every joint draw perpendiculars to their respective chords. Now the tread of the middle piece at *C* being just 8 steps, the height of the section from *h* to *m* is 8 steps; and the section *m n* is the same as *m n* on the falling mould, and the section *h i* is the same height as *h i* upon the falling mould; draw a line to touch the sections, and complete your face mould as in the foregoing plates: in each of the other pieces at *E* and *G*, the number of treads being 6, therefore, from your falling mould set the stretch of six steps; from *h* to H draw H *l*, parallel to *h n*, then H *k l* will give the height of the sections at *D* and *E*: everything else agreeably to the letters.

Note. The stretch-out of 8 steps, or any other number, is not reckoned on the chord; but it is the stretch-out round the convex side of the rail, or what most people call the inside.

Plate 69.

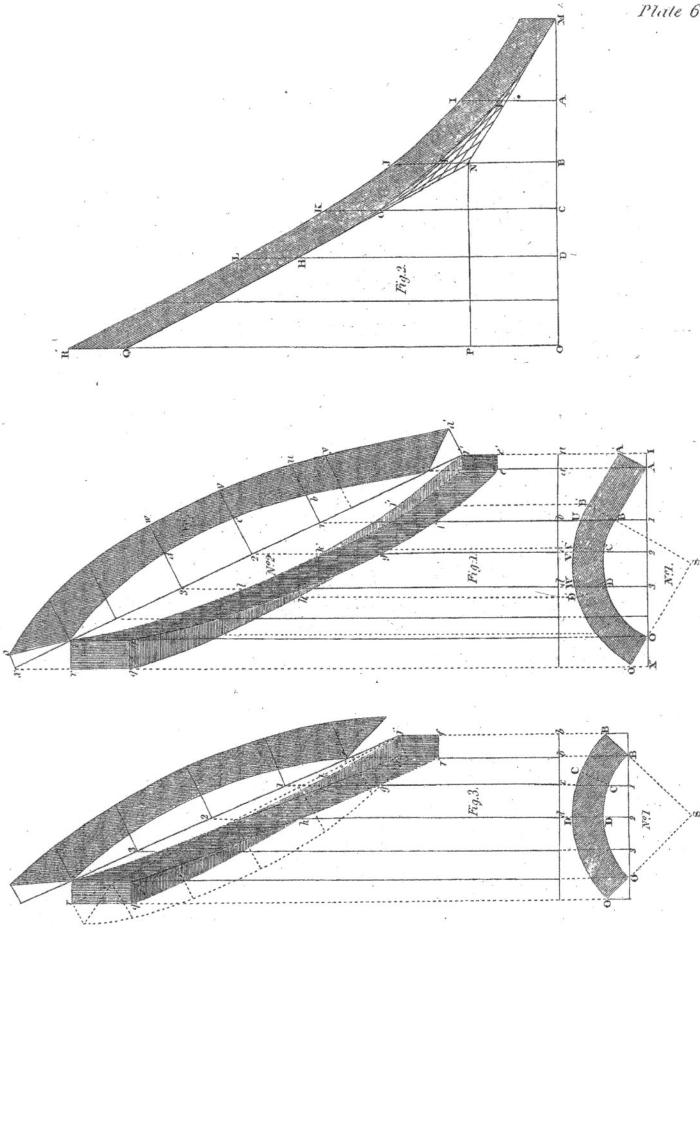

PLATE 69.

PART 1.

To show the proper twist of a rail and thickness of stuff, also the face moulds for the same.

FIG. 1. No. 1. *A A′ O′ O* be the plan of a rail: *figure* 2, the falling mould as completed. Let the plan of the rail No. 1, *fig.* 1, be so placed that the chord of *A C* be parallel to the base *M. O* of the falling mould *fig.* 2, and let *B B′* be the separation of the straight and circular parts of the rail, *O O′ B′ B* being the quadrantal part of the rail, and *B′ A′*, *A B* the straight part: divide the concave side *B C D O* of the quadrantal part into equal parts at the points *B, C, D,* &c., *O*; through all the points *B, C, D,* &c., draw lines at right angles to the chord *A O*, cutting it in 1, 2, 3, &c., and produce them upwards to the points *f g, h,* &c.; let *M B N*, *fig.* 2, be the section of a step, upon *M O* make *B A* equal to *B A*, No. 1, *figure* 1: extend the parts *A B, B C, C D,* &c. No. 1, *figure* 1, upon the base *M O*, *figure* 2, from *A* to *B*, from *B* to *C*, from *C* to *D*, &c.: from the points *A, B, C, D,* &c., draw lines perpendicular to *M O*, cutting the lower edge of the falling mould at *E, F, G, H,* &c., and the upper edge at *I, J, K, L,* &c. In *figure* 1, draw any line, *a, b, c, d,* &c., parallel to the chord *A O*; make *a e, b f, c g, d h,* &c., respectively equal to *A E, B F, C G, D H,* &c., *figure* 2; also in *figure* 1, make *a i, b j, c k, d l,* &c., equal to *A I, B J, C K, D L,* &c.: through the points *e, f, g, h,* &c., *figure* 1, draw a curve; also through the points *i, j, k, l,* &c., draw another curve, and these two curves will complete the projection of the falling mould. From the point *s*, *figure* 1, radiate the lines *C C′*, *D D′* &c., cutting the convex side at *C′ D′*, &c.: from the points *A′, B′, C′, D′,* &c., draw the lines *A′ i′, B′ j′, C′ k′, D′ h,′* &c.: draw *e e′, i i′, j j′, k k′, h h′,* &c., parallel to the chord *A O*: through the points *i, j, k,* &c., draw a curve until it intersect with the curve *i, j, k,* &c., this will form the projection of the top of the rail piece. In like manner the under parts which appear at *q q′, h h′,* will be completed in the same manner, so that No. 2 is the whole appearance of the solid, *q r′ r q′* and *i i′ e e′* being sections or the ends which join the contiguous parts of the rail.

PLATE 69.

PART 2.

To trace the face mould No. 3; join *i′ r*, and let the perpendiculars cut *i r* at *t*, 1, 2, 3, &c., draw the ordinates 1 *b u*, 2 *c v*, 3 *d w*, &c., perpendicular to *i r*: make 1 *b*, 2 *c*, 3 *d*, &c., equal 1 *B*, 2 *C*, 3 *D*, &c., No. 1: also in No. 3, 1 *u*, 2 *v*, 3 *w*, &c., equal to 1 *U*, also make *x o* equal to *X O′*, *i a′* equal to *I A′*, 2 *V*, 3 *W*, and *C*: draw *b y* parallel to *t a′*, and *a y* parallel to *t b*, and *t a′ y b* will complete the straight part of the face mould: join *r o*; draw the concave curve *b e d* . . . . *r*, and the convex curve *y u v w* . . . . . *o*, which completes the curve part and the whole of the face mould No. 3.

FIG. 3 shows the projection and face mould for the quadrantal part of a rail; the principles of projection and manner of drawing the face mould are the same as what have now been shown. This figure is introduced to show how much less wood the part of the rail requires from a quadrantal plan, than when a straight part of the rail is taken in, and the more of the straight rail that is taken in, the greater will be the deflection from the chord; thus in *figure* 3, the distance between the chord *j′ r* at any point to the nearest point of the projection, is much less than the distance *i′ r*, *fig.* 1, from a corresponding point in *i′ r* to the nearest point of the projection.

The dotted lines, *fig.* 3, show another face mould, the chord being drawn from the one extreme point to the other, of the upper end of each section, which makes the face mould much shorter than it ought to be, and will therefore require a much greater thickness of stuff. From this false position *j r* of the chord, some unskilful teachers of drawing direct their pupils to trace the face mould of a hand-rail.

Plate 70.

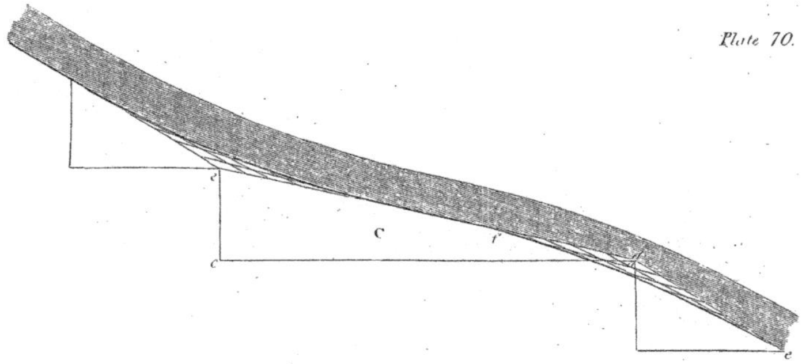

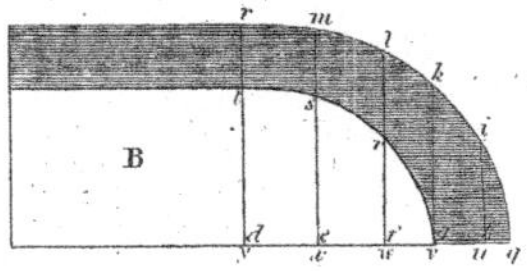

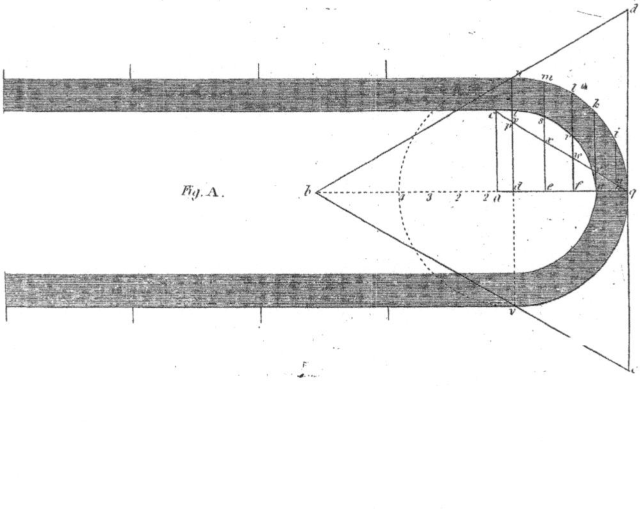

Fig. A.

PLATE 70.

To find the Face Mould of a Stair-case, so that when set to its proper Rake, it will be perpendicular to the plan whereon it stands for a level Landing.

In *fig. A* draw the central line *b q*, parallel to the sides of the rail; on the right line *b q* apply the pitch-board of a flyer; from *q* to *a* draw ordinates *n d*, *m e*, *bf k g*, and *i k*, at discretion, taking care that one of the lines, as *k g*, touch the inside of the rail at the point *g*, so that you may obtain the same point exactly in the face mould; then take the parts *q u*, *u v*, *v w*, *w x*, *x y*, and apply them at *B* from *q u*, *u v*, *v w*, *w x*, *x y*, from these points draw the ordinates of *B*, and prick them from the plan, *figure A*, as the letters explain; then *B* will be the mould required.

To find the Falling Mould.

Divide the radius of the circle *fig. A*, into four equal parts, and set three of these parts from 4 to *b*; through *n* and *v*, the extremities of the diameter of the rail, draw *b n* and *b v*, to cut the tangent line at the points *c* and *d*; then will *c d* be the circumference of the rail, which is applied from *c* to *d*, at *C*, as a base line: make *c e* the height of a step; draw the hypothenuse *e d*, at the point *e* and *d*; apply the pitchboard of a common step at each end of their bases, parallel to *c d*, make *d f* equal to *d e*, if it will admit of it, and by these lengths curve off the angles by the common method of intersecting lines; then draw a line parallel to it, for the upper edge of the mould.

PLATE 71.

To draw a Falling Mould for a Rail having Winders all round the circular Part, thence to find the Face Mould.

To describe every particular in this, would almost be repeating what has been already described; the heights are marked the same upon the falling mould at *D*, as they are at the face mould, which will give the heights of the sections of the rail; and the face mould at *C* is traced from the plan *B*, according to the letters; *G* shows the application of the mould to the plank; take the bevel at *H*, and apply it to the edge of the plank at *D*, and draw the line *b c*; then apply your mould to the top of the plank, keeping one corner of it to the point *b*, and the other corner close to the same edge of the plank; then draw the top face of the plank by your mould; then take your mould, and apply it to the under side at *c*, in the same manner.

Plate 71.

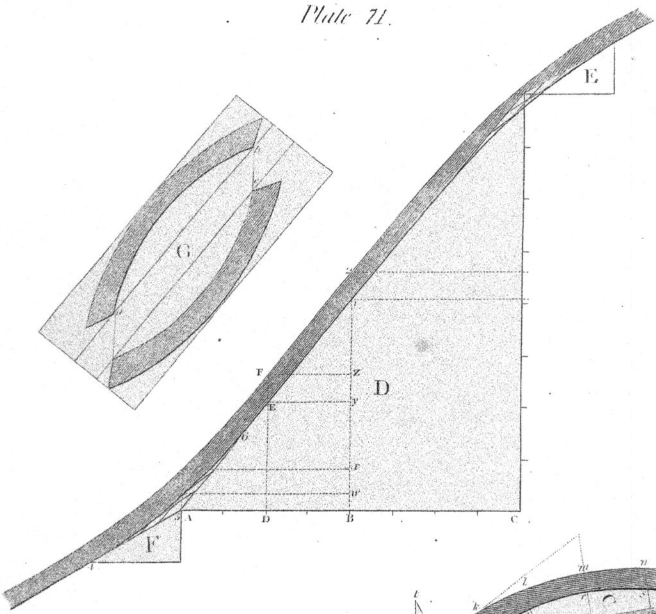

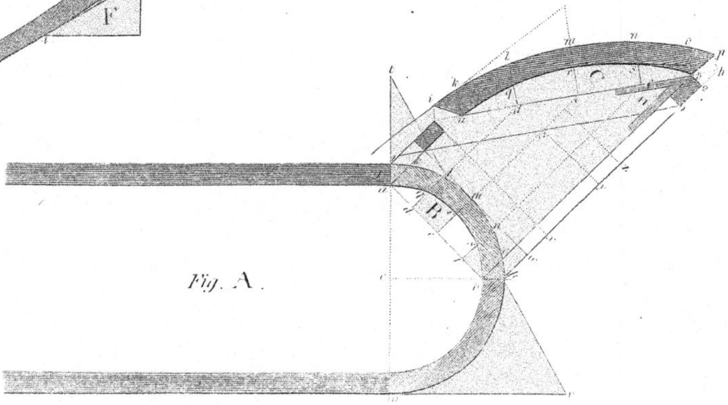

Fig. A.

Plate 72.

Fig. A.

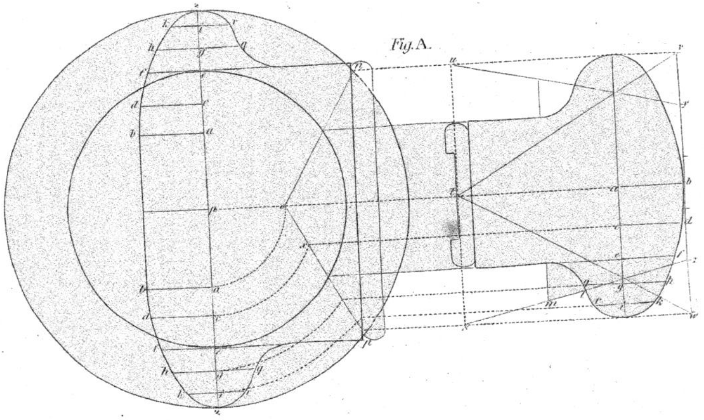

Fig. B.

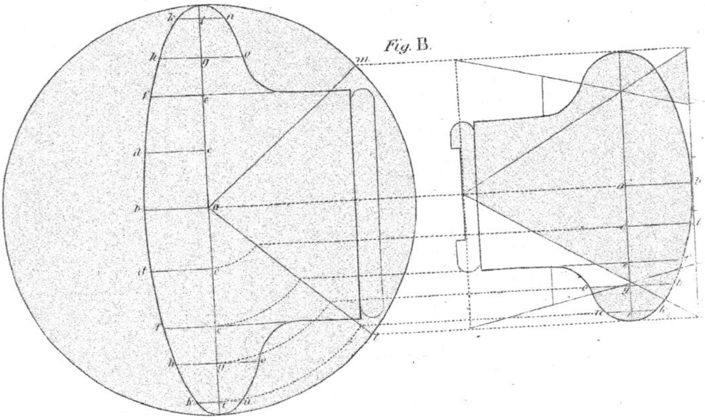

PLATE 72.

To draw the Form of a Hand-rail.

In *fig. A* make an equilateral triangle *v w t* upon its width, and divide it into five equal parts, and from one part on each side draw *z s* and *y w*, then *t*, *g* and *m* are the centres, *l m* being made equal to *l g*; the centres are found the same for the other side.

The Form of a Rail being given, to draw the Mitre Cap.

Let the projection of the cap be three inches and a half, and make the distance of the inside circle from the outside circle the projection of the nose on each side of the rail, and draw the mitre *n o* and *p o*; then continue parallel lines down to the mitre *p o*; put the foot of your compass in the centre of the cap, and circle the parallel lines round to *a*, *c*, *e*, *g*, and *i*, and draw the ordinates *a b*, *c d*, *e f*, &c., and then prick the cap to the rail according to the letters.

How to draw the Form of the Cap for the Mitre to come to the Centre.

It is only drawing the parallel lines from the rail to the mitre wherever it is, and circling them round to the ordinates, and so pricked from the rail, and the thing is done.

PLATE 73.

How to diminish the Shaft of a Column by the ancient Method.

In *fig. A*, describe a semicircle at the bottom; let a line be drawn through the diameter at the top, parallel with the axis of the column, till it intersects the semicircle at 1, at the bottom; then 1, 1 at the bottom will be equal to 1, 1 at the top; divide the arch into four equal parts, and through these points draw lines parallel to the base, the height of the column being also divided into the same number of parts, and lines drawn parallel to the base, then the column is to be traced from the semicircle, according to the figures.

How to diminish the Column by Lines drawn from a Centre at a Distance.

FIG. *C.* Take the diameter *a b*, at the bottom, set the foot of your compass in *c* at the top, and cross the axis in the point *d*, continue *c d* at the top, and *a b* at the bottom, to meet at *e*; then draw from *e* as many lines across the column as you please, and take the diameter *a b* at the bottom, and prick each line upon the axis equal to *b a*, which will give the swell of the column.

To diminish a Column by Laths, upon the same Principle.

In *fig. D*, the point *e* being found, as in *fig. C*, take and plow a rod *d b*, and lay the groove upon the axis of the column, and plow the describing rod upon the under side, and lay the groove upon a pin fixed at *e*, and fix a pin at *g*, to run in the groove upon the axis of the column, and the distance of the pencil at *f*, equal to *b a*, then move the pencil at *f*, it will describe the diminishing.

How to describe the Column by another Method.

Take the semi-diameter *a b* at bottom, and set the foot of your compass in the top at *c*, and cross the axis at 8, and draw the line *a* 8 on the outside, parallel to *b* 8 on the axis, and divide each of these lines into eight equal parts, and set the diameter *a b* at the bottom along the slant lines 1 1, 2 2, 3 3, &c., from the axis; this will also give the diminishing of the column.

How to make a diminishing Rule.

Divide the height of your rule into any number of equal parts, as 6; draw lines at right angles from these points across the rule, and divide the projection of the rule at the top; that is, half of what the column diminishes; into the same number of equal parts put a pin or brad-awl; lay a ruler from *a* to 5; mark the cross line at *f*; then lay a ruler from 4 to *a*, and mark the next cross line at *e*; then lay the ruler from 3 to *a*, mark the next at *d*, and so on to the bottom; bend a slip round these points, and draw the curve by it, will give a proper curve for the side of the column.

Note. This is the readiest method, and gives the best curve of any that I have tried.

Plate 73

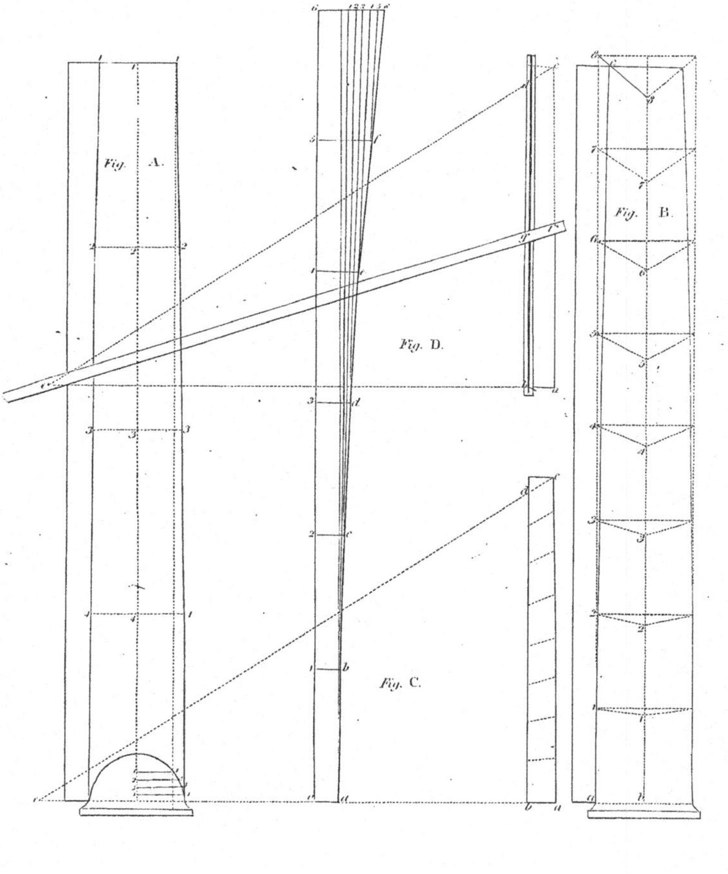

Plate 74.

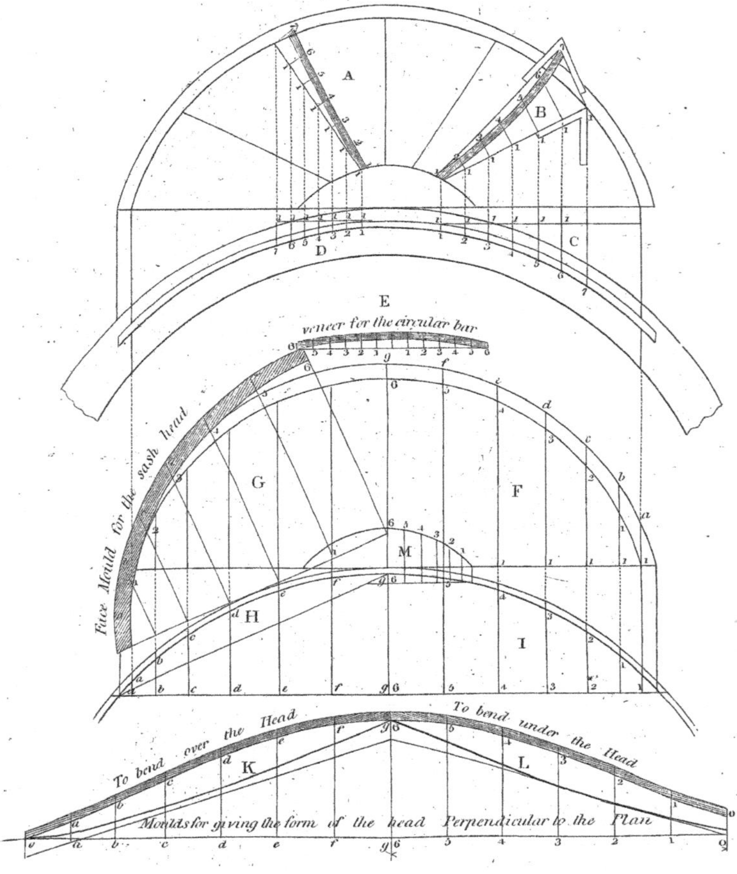

A
B
C
D
E
veneer for the circular bar
Face Mould for the sash head
G
F
M
H
I
To bend over the Head
To bend under the Head
K
L
Moulds for giving the form of the head Perpendicular to the Plan

PLATE 74.

The Plan and Elevation of a circular Sash, in a circular Wall, being given; to find the Mould for the radial Bars, so that they shall be perpendicular to the Plan.

Draw perpendiculars from the points 1, 1, 1, 1, &c., at *A* and *B*, in the radial bars, either equally divided, or taken at discretion down upon the plan, to 1, 2, 3, 4, 5, 6, 7, at *C* and *D*; and draw a line from the first division upon the convex side parallel to the base; then draw ordinates from 1, 1, 1, 1, &c., at right angles to the radial bars, at *A* and *B*, which being pricked from the plans at *D* and *C*, will give a mould for each bar; and the bevels upon the end will show the application of the moulds.

To find the Veneer of the Arch-bar, or what is improperly called by Workmen Cot or Cod-bar.

To avoid confusion, I have laid down the plan and elevation for the head of the sash under. The stretch-out of the veneer is got round, 1, 2, 3, 4, 5, 6, on the arch-bar, which being pricked from the small distance on the plan at *M*, will give the veneer above, at *E*.

To find the Face-mould for the Sash-head.

Divide the sash-head round, into any number of equal parts, at *G*, and draw them perpendicular to the base at *H*; draw the chord of one-half of the plan at *H*, and draw a line parallel to it to touch the plan upon the back side; then the distance between these lines at *H*, will show what thickness of stuff the head is to be made out of; and from the intersecting points on the back side, draw perpendiculars from the base of the face mould, which being pricked from the elevation, as the figures direct, will give the face mould.

To find the Moulds for giving the Form of the Head, perpendicular to the Plan.

The base of *L* is got round the arch 1 2 3 4 5 6, at *F*, and the base of *K* is got round *a b c d e f g*, also at *F*, and the heights of the ordinates of each are pricked either from *H* or *I*, which will give both moulds.

By the same method, a circular achitrave, in a circular wall, may be got out of the solid.

Note. The face mould at *G* must be applied in the same manner as in groins; so that the sash-head must be bevelled by shifting the mould *G*, on each side, before you can apply the moulds *K* and *L*; the black lines at *K* and *L* are pricked from the plan, at *H*; these black lines will exactly coincide with the front of the rib when bent round; a line being drawn by the other edge of the moulds, will be perpendicular over its plan, and the thickness of the sash-frame towards the inside will be found near enough by gauging from the outside.

PLATE 75.

Construction of a circular headed architrave in a circular wall. See the description on the plate.

Plate 75.

An Architrave or Archivolt for a Circular Window in a Circular Wall

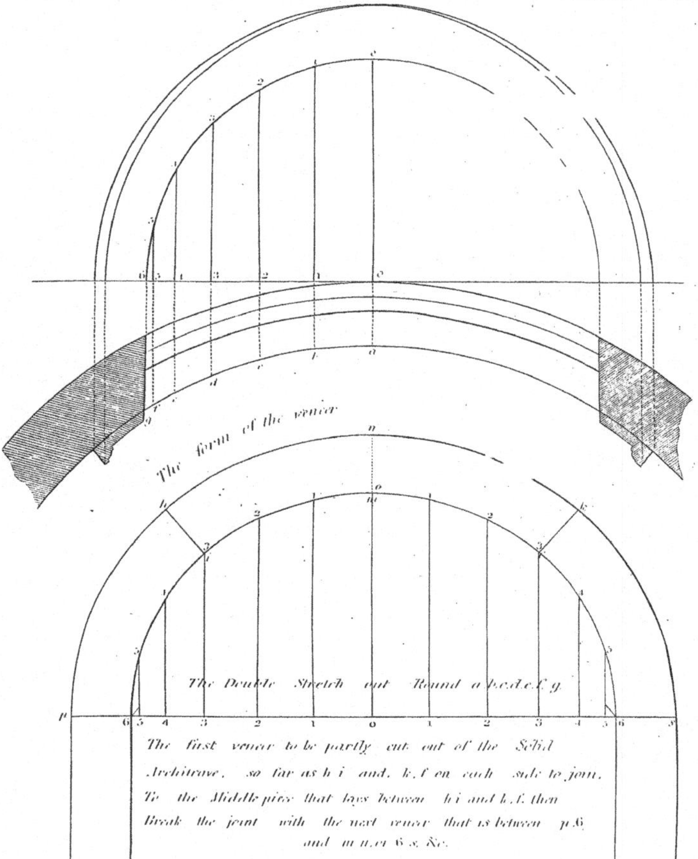

Plate 76.

The Method of getting out angle bars for Shop Fronts.

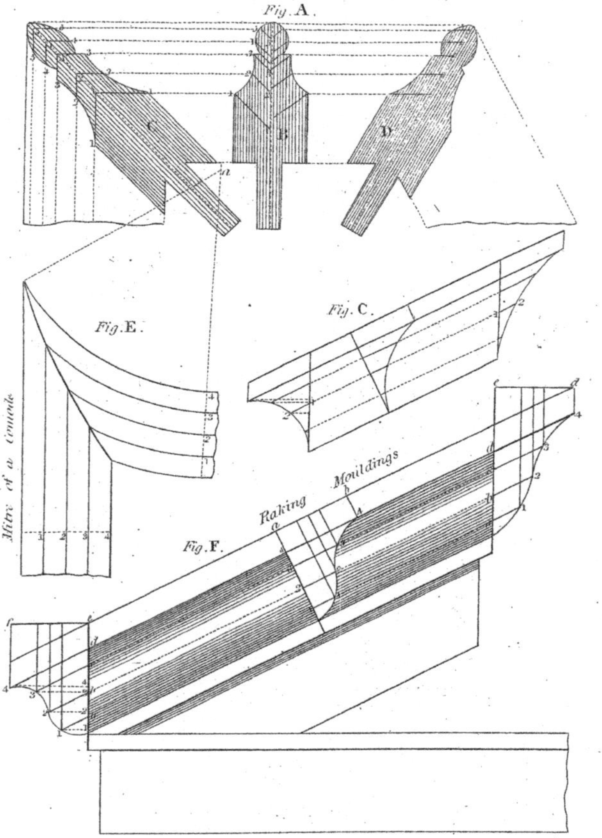

PLATE 76.

To describe the Angle Bars for Shop Fronts.

In *fig. A, B* is a common bar, and *C* is the angle bar of the same thickness; take the raking projection 1, 1, in *C*, and set the foot of your compass in 1 at *B*, and cross the middle of the base at the other 1; then draw the lines 2, 2; 3, 3, &c., parallel to 1, 1; then prick your bar at *C* from the ordinates so drawn at *B*, which being traced will give the angle bar.

How to draw the Mitre Angle of a Commode Front for a Shop.

In *E* divide the projection each way in a like number of equal parts, then the parallel lines continued each way will give the mitre.

How to find the Raking Mouldings of a Pediment.

In *fig. F* let the simarecta on the under side be the given moulding, and let lines be drawn upon the rake at discretion; but if you please, let them be equally divided upon the simarecta, and drawn parallel to the rake; then the mould at the middle being pricked off from these level lines at the bottom, will give the form of the face. The return moulding at the top must be pricked upon the rake, according to the letters.

The cavetto, *fig. C*, is drawn in the same manner.

N. B. If the middle moulding, *fig. F*, be given, perpendiculars must be drawn to the top of it; then horizontal lines must be drawn over the mouldings at each end, with the same divisions as are over the mouldings; and lines being drawn perpendicularly down, as above, will show how to trace the end mouldings.

PLATE 77.

Raking Mouldings, and Mouldings of different Projections.

FIGURES *B* and *A* show how to trace base mouldings for skirting to stairs, upon the same principles as shown in the last plate; at the bottom are given two methods of mitring mouldings of different projections together.

Plate 7

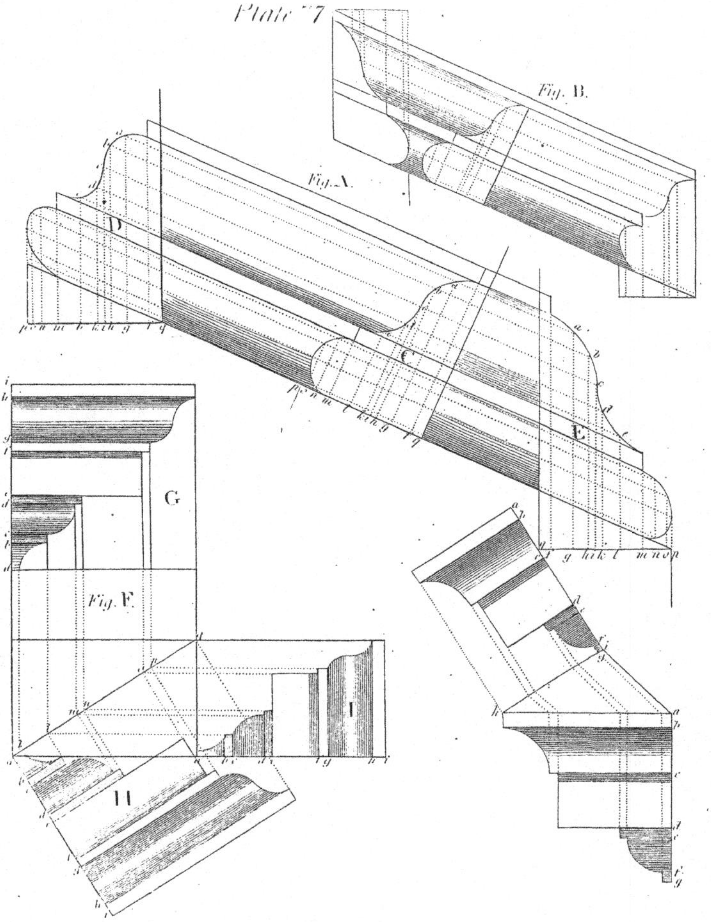

Plate 78.

This shows the method of enlarging cornices

Fig. A.

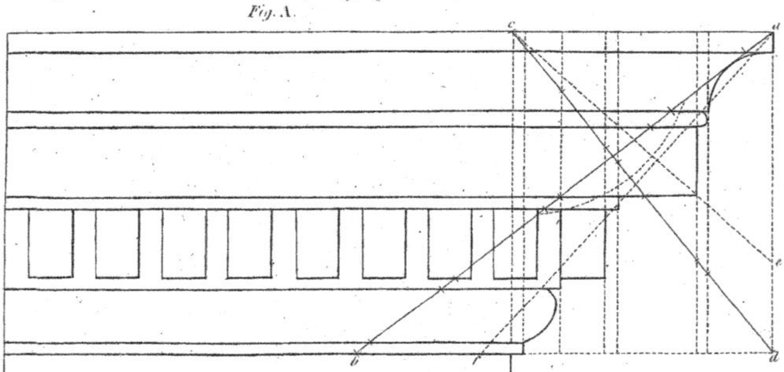

This shows the method of diminishing cornices

Fig. B.

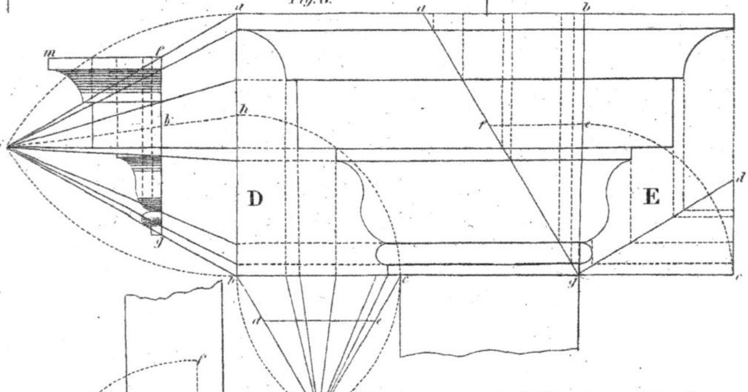

This shows the method of hanging a jib door

Fig. C.

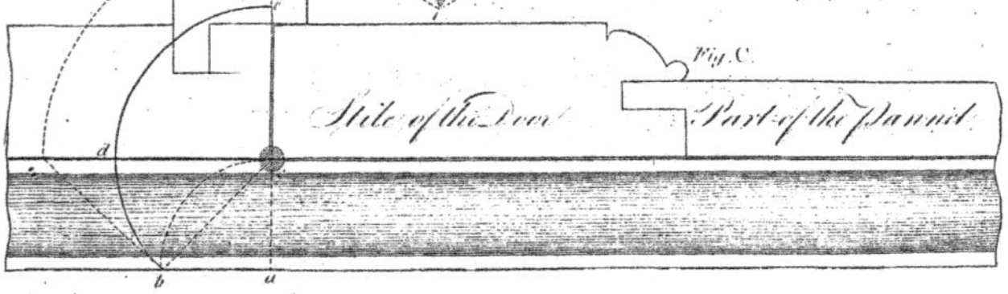

PLATE 78.

Given the Form of a Cornice, to draw it to a greater Proportion.

In *fig. A*, let the given height of the cornice be *a d*; set one foot of your compass in *a*, and cross the under side at *b* with that height, and from the point *c* draw the line *c d* at right angles to *a b*; then the height of all your mouldings will be the parts of *a b*, and the projections the parts of *c d* in proportion.

Note: *a f* shows another height, *c e* its projection in proportion to that height.

How to diminish a Cornice in the Proportion of a greater.

Describe equilateral triangles on the base and projection as at *D*, and make *i f* and *i g* equal to the intended height, and draw the line *f g* across the triangle, which will give the heights in proportion to *a b*; put the foot of your compass in *b* as a centre, and circle *b c* round *b h*, and draw the dotted line *h i*, cutting *f g* in *k*; then set off *i e* and *i d*, each equal to *g k*; draw *d e*; then take the divisions of *e d*, and set them from *f* to *m*; in the same order draw perpendiculars: it will give the diminished cornice at *D*.

Another Method.

At *E*, let the given height be *a b*, and draw the hypothenuse *a g*, and lines being squared up to *a b*, from the divisions of *a g*, will give the heights; and if you draw the line *g d* at a right angle with *a g*, then *d c* will give the projection in proportion, when returning upon *d c*.

FIG. C *is the Method for hanging a Jib Door.*

Let *a c* be the projection of the surbase or base moulding, and *c* the centre of the hinge; make *a b* equal to *a c*, and in the centre at *c* describe the arch *b d e*; then the arch *b d e* will be the proper joint for the surbase to work in.

The joint of the surbase or the base may also be straight, as you see by the dotted line touching the circle at the point *b*, as the tangent to it.

PLATE 79.

MOULDINGS UPON THE SPRING.

To find the Sweep of a Moulding to be bent upon the Spring round a circular Cylinder.

In *fig. A*, which stands upon a semicircular plan, make *a c* equal to the height of your moulding, and make *a b* equal to the projection; describe the form of the moulding, and draw a dotted line to touch the face of it; then draw the line *e d* to meet in the centre of the body, at *d*, so as to keep your moulding to a sufficient parallel thickness; from the centre *d* describe the several concentric circles which are the arrises of the moulding required.

How to find the Sweep of your Moulding when the Plan is a Segment.

Complete the semicircle as in Plate 1, *fig.* 1, then proceed as described in *fig. A*.

FIGURES *C* and *D* show the method for bending a moulding round the inside, which is performed the same as above.

The demonstration may easily be conceived from the covering of a cone.

Plate 79

Fig. A.

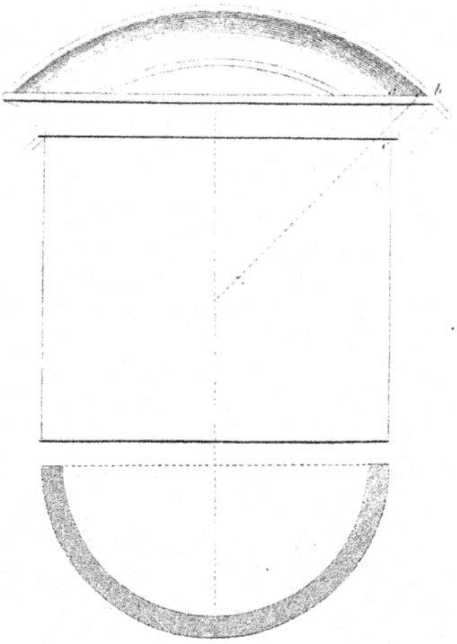

Fig. B.

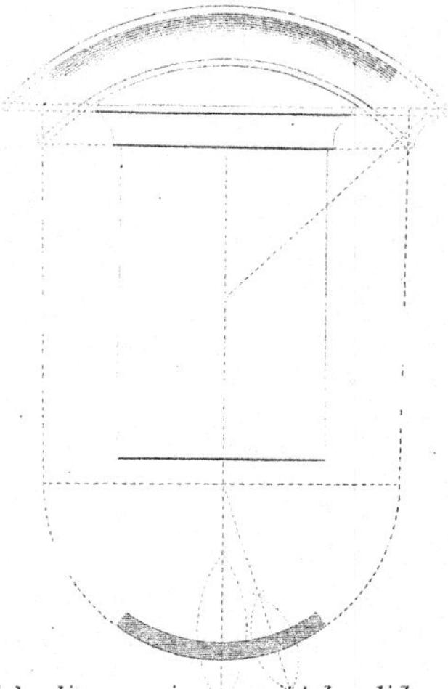

This Plate shows the different methods of bending cornices cut of the solid to stand to any Spring.

Fig. C.

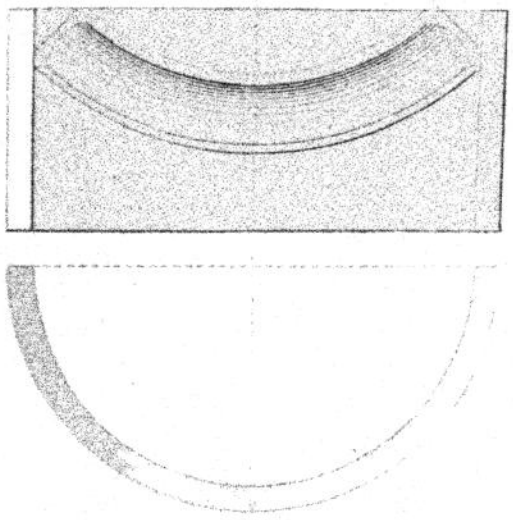

Fig. D.

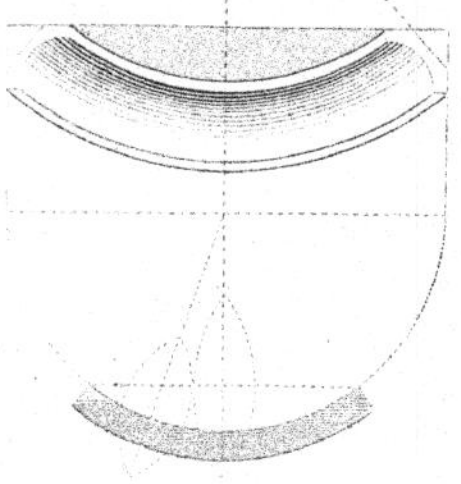

Plate 80.

Fig. A.

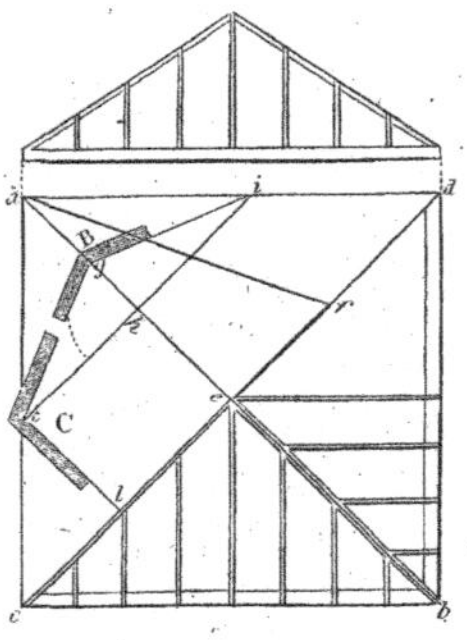

Fig. B.

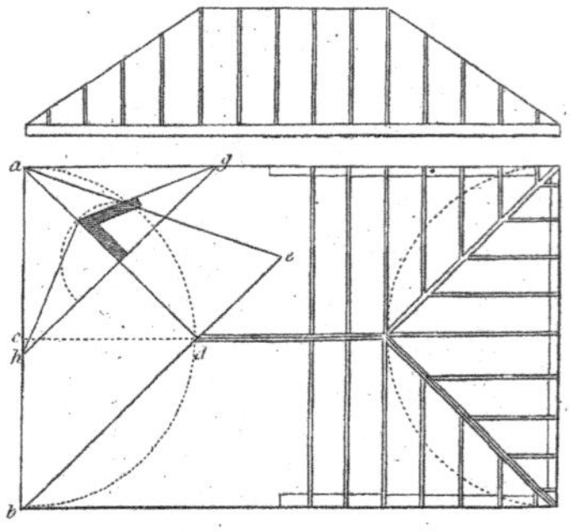

Fig. C.

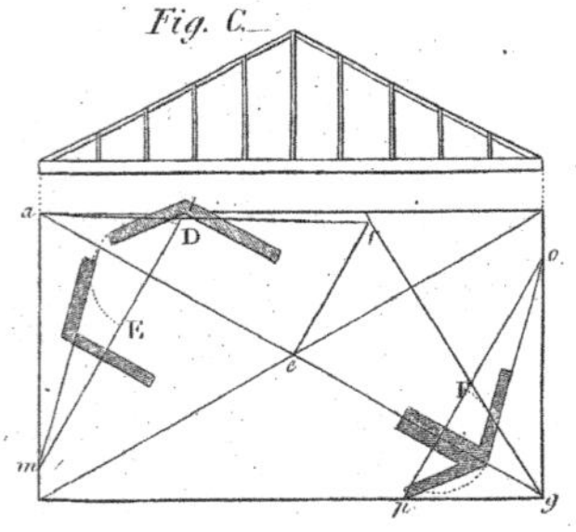

Fig. D.

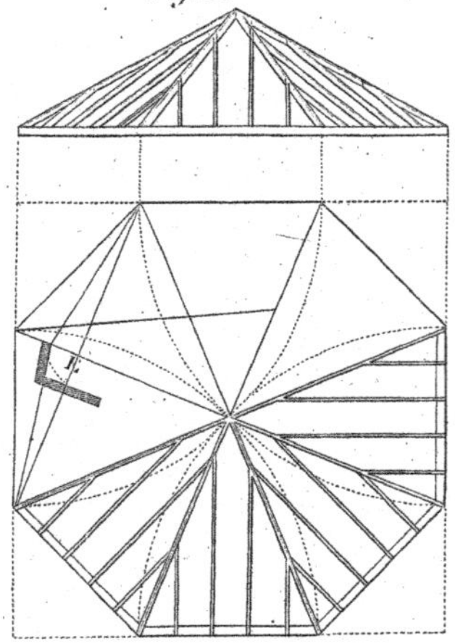

Fig. E.

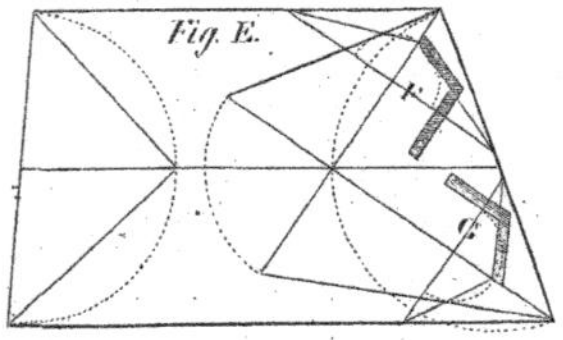

PLATE 80.

To find the Length of the Hips of a Sky-light standing upon a square Plan, the Height being given.

In *fig. A*, draw the diagonals *a b*, and *c d*; they will bisect each other at right angles at *e*; take *a e* for the base of any hip; from *e* in *e d* make *e f* equal to the height of the sky-light; join *a*, *f*, and *a f* will be the length of the hip required.

To find the Backing of the Hip.

Draw any line *k i*, at right angles to *a e*, the base of the hip rafter, cutting it in any point *h*, put the foot of your compass in *h*, as a centre, and with the other describe a circle to touch *a f*, the hip rafter, to cut the base line *a e*, at *g*; then draw *g i* and *g k*; then the angle *k g i* will be the backing of the hip, as is shown by the bevel at *B*; but the best way to work the hips is to apply a bevel to the parallel sides of the hips, as is shown at *C*, by making the other side of the bevel parallel to *a e*, the base of the hip.

Note. The same lines will extend to any sky-light, whatever may be the form of its plan; if it be any polygon, to find the length of the hip rafter, draw a line through any point in its base at right angles to it, so as to cut the two contiguous sides to that base, and on the said point as a centre describe a circle to touch the hip rafter from the points where this circle cuts the base line, draw two lines to meet the ends of the perpendicular line at the sides of the polygon; then the angle formed by these two lines will be the backing required: but perhaps a few more examples will make it plainer than many words can.

Fig. *B* is a sky-light, standing upon a rectangular base, having a ridge in the middle; make *c d* upon the ridge line equal to half the width of *a b*; then the angle *b d a* will be a right angle: every other requisite is the same as directed for *fig. A.* If these hips are to be mitred, the bevel at *C* shows the mitre.

Fig. *C* is another sky-light, standing also upon a rectangular base; but the hips all meet over the centre of the plan at *e*, and consequently the diagonals do not bisect each other at right angles; therefore take any base line, as *a e*, or *e g*, and make *e f* perpendicular to *a e*, from *e*, and equal to the height of the sky-light; and draw *f a* or *f g*, for the length of the hip, by drawing the line *l m* at right angles to *a e*. The backing will be found in the same manner as the others above. This sky-light will require two different bevels *D* and *E*, to be applied to the parallel sides of the hip, which are both found from the backing by drawing the stocks of the bevel parallel to *a e*, the base of the hip.

But if the hips are to be mitred together, *F* and *G* show the two bevels for the mitring each half, so that when put together shall form the proper backing.

Fig. *D* is a sky-light standing upon an octangular plan, as is described in *fig.* 8, Plate 6, of the Geometry; the lengths of the hips and backing of the angle are found in the same manner as directed for others.

Fig. *E* is a sky-light whose plan is trapezoid; upon each end as a diameter describe a semicircle to cut the ridge line; from these points draw lines to the extremities of their respective diameters, which will form a right angle for the base of the hips to stand upon; the backing or mitring of the hips will be found as is described in *fig. A* and *B*.

Plate 81

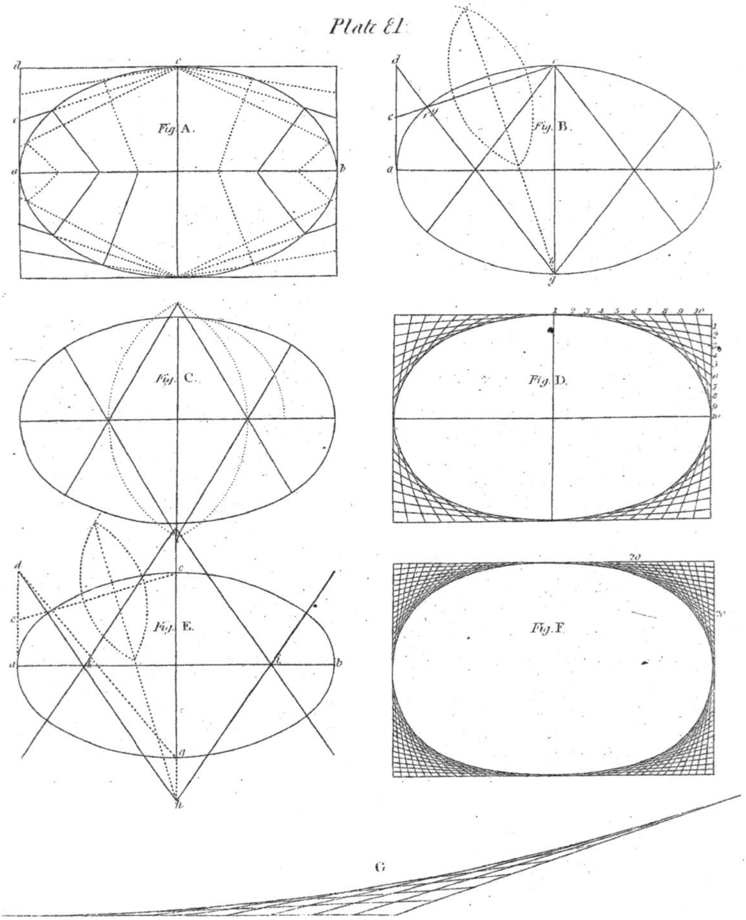

CONCLUSION.

In which are examined, by way of preventive Caution to the Student, several Methods, which are founded on wrong Principles, and better ones are here proposed.

PLATE 81.

Of the Ellipsis.

THE old manner for intersecting all kinds of lines, applied to Gothic and elliptical figures, to this day is exceedingly useful in forming the ramps of stairs, or easing of any angle, as at *G*; but when this is applied to elliptical figures, it is far from forming a true ellipsis, being too full at the ends, as at *fig. D* and *E*; and this is no certain rule for drawing an ellipsis, for the more divisions there are, the worse is the ellipsis; as for example, *fig. F* is divided into double the number of parts as *D*; it is plain that neither *D* nor *C* is an agreeable ellipsis, and *C* is much worse than *D*, which is contrary to general opinion; for I have been frequently told, the more parts it is divided into, the truer it is: but by this it appears the more parts it is divided into, the worse it is: if this is doubted, try. *Fig. A* is an ellipsis drawn on true principles, as laid down in Plate 7, at *C*, of this book, and is here repeated to be compared with the others. *Figures B, C,* and *E,* are ellipses drawn with a compass: I may call them representations, as it is impossible to draw them true with a compass; there is no part of the curvature of an ellipsis that will exactly agree with any part of a circle, for in every quarter of an ellipsis, from the extremity of the transverse, the curvature in every succeeding part is continually flatter towards the extremity of the conjugate axis; but yet there is a method to represent an ellipsis, which will differ very little from the truth, as is shown at *figures B* and *E*, which are both drawn by the same method; *fig. E* or *B* is the nearest to the shape of *fig. A*; *fig. C*, the method used by almost every author who has written upon the subject, as full at the ends, but not in so great a degree as *fig. D* and *E*.

PLATE 82.

Of Raking Mouldings.

In Plate 76, *fig. A*, let the moulding at the bottom be given, and let the perpendicular height be divided into any number of equal parts at 6; likewise divide the perpendicular height of the top moulding into six; and the face moulding into six, at right angles to the rake; and let the ordinates of each be drawn through the equal divisions of each respective perpendicular, and pricked from the bottom, as the figures direct. It is evident that if the under moulding is composed of two quarters of a circle, the upper mouldings will be composed of two quarters of an ellipsis; consequently, the return moulding at the top will be too quick upon the round, and likewise in the hollow.

But if this demonstration should not be sufficient, let us suppose a plane parallel to the arrises of the moulding, and perpendicular to the plane of projection, to pass through the point 1, this plane will be represented by a straight line; therefore let a dotted line be continued from 1, in the given moulding, parallel to the rake; then it will be evidently seen, that this dotted line corresponds with neither the face nor the return moulding; for in the face moulding it falls between the points 1 and 2; and in the top moulding it falls almost at the point 2; whereas it should only come to the point 1 in each; but the horizontal projection from 2 to 2, at the top, ought to be equal to 1 1 at the bottom; but it is much greater; therefore this method is false, and they will not mitre together.

I shall also notice another method used by some authors, see *figure B*, at *D* and *E*, where they are pricked perpendicular to their chords, in the middle, which is also false; but if they are pricked as at *B* and *C*, on the rake, will be exceedingly near, if described with a compass through three points.

Plate 82

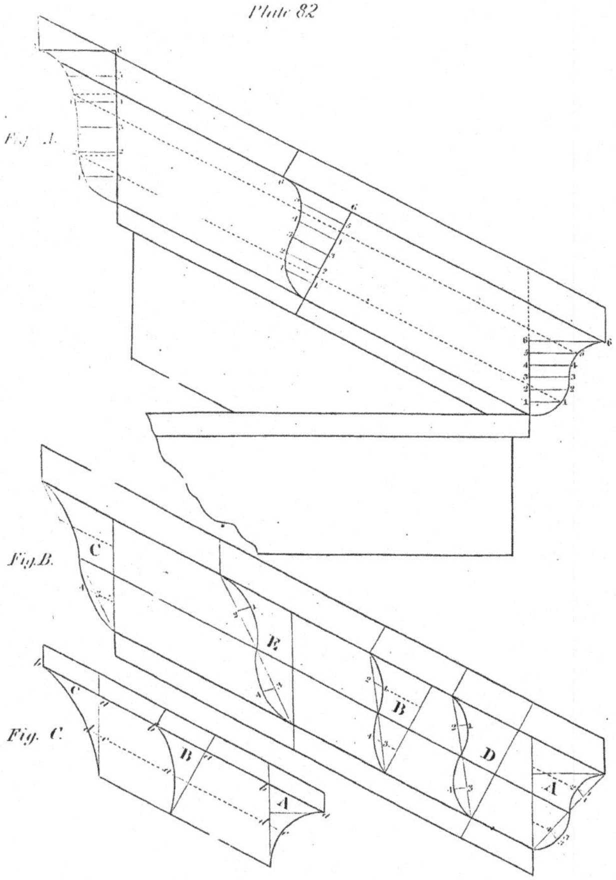

Plate 83.

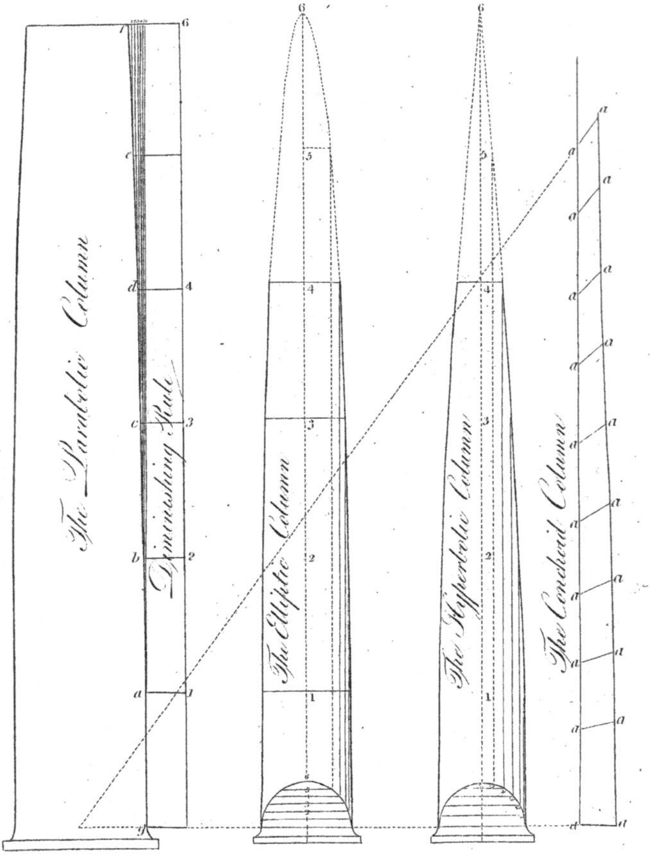
The Parabolic Column
Diminishing Rule
The Elliptic Column
The Hyperbolic Column
The Conchoid Column

PLATE 83.

Of diminishing of Columns.

The method for drawing a column, described by some authors, and which is properly called a conchoid column, is not only very inconvenient on account of the cumbersome instrument which is necessary to find the curve for practice, but also the appearance is, I think, less graceful than when produced by other methods. The conchoid curve, or column, is concave towards the bottom, and convex towards the top; and if this curve was infinitely extended, it would never meet the axis; which shows it to be different from the elliptic curve or column, as some have called it.

The column called hyperbolic, *Plate* 18, is not so named from the general properties of the conic hyperbola (because there may be an infinite number of hyperbolas standing upon the same base, having one common vertex, which will all be contained between a triangle and a parabola, according as its axis is longer or shorter), but because it will nearly coincide with some of these hyperbolas. This curve has been known among workmen, and by them has been mistaken for an elliptic curve; to refute which I have, on the same plate, shown a true elliptic column for comparison; the lines of their curvature are continued only to show their true figure; either of these is a more commodious method than the conchoid.

The method which I recommend as easiest in practice for diminishing of columns is already described on *Plate* 73, by means of a diminishing rule, which is infinitely more convenient than the trammel, and which to my eye, also produces a pleasant contour: but as this will depend on the fancy of the architect, the workman will find some of the methods shown will answer his purpose for any curve. The conchoid is flattest at the top, the hyperbolic is a little quicker, the parabolic is still more so, and the elliptic is the most quick.

INDEX.

Spotted something that isn't quite right?

Let us know and we'll fix it for future reprints.
And as a thank you for your diligence, we'll
send you a free Sothis book of your choice.

publisher@sothis.press

This page intentionally left blank

Except of course for that line above. And this one.
And whilst we have your attention, perhaps you'd
like to follow us on Instagram @sothis.press. Go on.

Made in the USA
Monee, IL
08 November 2026

68113588R00173